Building Planet Earth

Building Planet Earth presents a description of Earth as a planet, commencing with its physical and chemical evolution out of the primordial solar nebula. The condensation of elements and their redistribution are described, leading into a section dealing with mapping, geophysical and geochemical studies. This establishes the gross structure of the Earth, following which basic principles and processes of plate tectonics are then described, leading to the elucidation of the working of geological cycles. The main thrust of the remainder of the book is a description of the geological evolution of the Earth. Volcanism and seismicity, ice ages and climate, isotopic techniques and age dating, are all treated. The impact of mass extinctions, global-warming and ozone holes are included. The book is illustrated profusely and closes with a number of useful appendices.

PETER CATTERMOLE was formerly a lecturer in both terrestrial and planetary geology at the University of Sheffield, with research interests in volcanology and planetary geology. During his University tenure he was a Principal Investigator with NASA's Planetary Geology and Geophysics Program (studying volcanicity on Mars), a Leverhulme Research Fellow (working on volcanology in Indonesia) and a recipient of a Royal Society Fellowship for work on terrestrial analogs to Ionian volcanism. Since leaving the University fold he has concentrated on writing, lecturing and organising and leading specialist geological tours for Journeys of Special Scientific Interest, a small company he set up in 1992.

Beehive erosion of Devonian sandstones at Echidna Chasm, Bungle Bungle National Park, East Kimberley Plateau, Western Australia.

Building Planet Earth

Peter Cattermole

CAMBRIDGE
UNIVERSITY PRESS

NY   2000

PUBLISHED BY THE PRESS SYNDICATE OF THE UNIVERSITY OF CAMBRIDGE
The Pitt Building, Trumpington Street, Cambridge, United Kingdom

CAMBRIDGE UNIVERSITY PRESS
The Edinburgh Building, Cambridge CB2 2RU, UK   www.cup.cam.ac.uk
40 West 20th Street, New York, NY 10011-4211, USA   www.cup.org
10 Stamford Road, Oakleigh, Melbourne 3166, Australia
Ruiz de Alarcón 13, 28014 Madrid, Spain

First published 2000

Printed in the United Kingdom at the University Press, Cambridge

Typeset in Swift 9.25/14 pt. in QuarkXPress® [HM]

A catalogue record for this book is available from the British Library

Library of Congress Cataloguing in Publication data

Cattermole, Peter John.
Building Planet Earth/Peter Cattermole:
     p.   cm.
Includes bibliographical references and index.
ISBN 0 521 58278 4 (hc.)
1. Geology. 2. Earth. I. Title.
QE26 2.C384 2000
550–dc21   98-36460 CIP

ISBN 0 521 58278 4 hardback

Contents

Foreword

Geology is of interest to everyone, and it is all around us; it is the study of our planet in all its aspects, and, after all, Earth is the only world in our part of the universe which is suited to our kind of life. Yet many people have only a vague idea of what geology is all about, and most books are either too elementary or else too technical. Many problems have recently come to the fore; for example, are we responsible for global warming? Is there any danger from the depletion of the ozone layer above our heads? Can we make any progress in predicting violent earthquakes? Peter Cattermole deals with these and many other aspects – and does so in a way which will fascinate the beginner, with no previous scientific knowledge, and will also be of real value to the serious student and researcher.

This is a splendid overall survey, written by a leading expert in the subject. *Building Planet Earth* fills a notable gap in the literature; it deserves, and I know will have, a widespread and lasting readership.

PATRICK MOORE

Preface

It is over a decade since *The Story of the Earth* was written, in collaboration with my long-standing friend, Patrick Moore. He has decided to stand down from being involved with this new book, but very kindly has provided it with a foreword – for which I am very grateful. It was he who first suggested that I should become a geologist.

The Earth has continued to go around the Sun and the Sun, perforce, moves around the Universe. The same physical laws still govern what happens in and around the Earth, although our views of some of the processes which operate to modify the crust, oceans and atmosphere have changed. Furthermore, research undertaken over the past ten years or so has added greatly to our knowledge of the details of geology and of geophysical processes. New ideas have arisen and theories which were new at the time of the first edition have been tested. Thus it is that science progresses.

Man's influence on the natural environment has been closely monitered in the last decade. Levels of pollution generally have risen on the global scale, as industrialization reaches the Third World. The motor car continues to seduce mankind, generating in turn much filth which rises into the atmosphere. The use of chlorine-based chemicals has contributed to an ever-enlarging hole in the Earth's protective ozone layer. At present the biggest ozone deficit is to be found over Antarctica, but a smaller hole has been detected at the antipode. What short- and long-term effects mankind is having on the global climate is not entirely clear. That changes have occured cannot be disputed, but whether these have been catalysed by human activities, or would have happened anyway, we cannot be sure. Much doom-and-gloom has been presented by the media but causes and effects are not always simply connected. The Earth and its atmosphere, hydrosphere and biosphere constitutes an incredibly complex natural system.

Building Planet Earth retains much of the material which resided within The Story of Earth's pages; however, substantial revisions have been made and much new material has been added. Since 1985 many volcanoes have violently erupted, there have been numerous earthquakes, and the lithospheric slabs which carry us all along have moved slowly but inexorably along on the sluggish mantle beneath. Herein I report some of the recent events and hope to have been successful in presenting a balanced account of how our planet works, and the sequence of its history. I would like to tender my thanks to Dr Simon Mitton and Cambridge University Press for affording me the opportunity – gladly grasped – to revise and update the original work. It has been a pleasure for me and I hope that the reader may derive some measure of my own enjoyment from the finished product.

Finally, the original artwork for *Story of the Earth* was drawn by Paul Doherty who, sadly, passed from this world at the tender age of 49 years. I would like to dedicate this book to his memory.

PETER CATTERMOLE
Sheffield
December 1999

Part one
Beginnings

Sandstone spires etched out of the Permian-age Coconino Sandstone, Sedona, Arizona. These colourful red and orange sandstones were laid down by the wind in ancient deserts that covered this part North America 250 million years ago.

1

Planet Earth

Introduction

In the pages which follow I will endeavour to tell the story of the Earth, from its birth some 4600 million years ago until the present day. Geology is the study of the Earth, but in recent decades, as mankind has sent spacecraft to all of the planets, save Pluto, its scope has been extended to encompass other planets and their moons; for this reason the term earth science is often used to define geology as traditionally conceived. Several branches of science, including physics, chemistry, meteorology, biology and astronomy have to be drawn upon to unravel the complex sequence of events that contributed to the Earth's geological story.

Ancient civilizations believed Earth to be flat, with the heavens revolving around it once every day; even when the Greeks showed it to be a globe, it was still thought to be the most important body in the Universe. This was natural enough, and the idea that the Earth was supreme persisted for a surprisingly long time. As recently as 1600 the Italian philosopher Giordano Bruno was burned at the stake in Rome, one of his heretical beliefs being that the Earth moves around the Sun. The final proof that it does, and that it is but an ordinary planet, did not come until the seventeenth century. We now know that all of the planets were formed at the same time as the Sun which occupies a position central to the Solar System.

The problem of how life began is also a part of Earth's story. It is generally believed that life began on the Earth itself, some time after the formation of the planets, perhaps around 3000 million years ago, but it has also been suggested that life was implanted from outer space, probably by passing comets. Whether or not this is true we do not know, neither have we definite proof that life exists elsewhere. The likelihood that a subcrustal water ocean exists on Jupiter's moon, Europa (and might provide a suitable environment for the development of some kind of life), and the discovery of bacteria-like particles within an Antarctic meteorite that many believe to have originated on Mars, are two recent discoveries which have relevance to this topic. The other planets in the Sun's family are unsuitable for the development of life in their various ways, but the Sun is only one of a hundred thousand million stars in our local Galaxy – and the Galaxy itself is one of many. It seems both illogical and conceited to assume that mankind is unique.

Many of the geological problems which seemed baffling not so very long ago have now been solved. We have a pretty good idea as to how our planet formed, and we can fix its age with a fair degree of accuracy; the record of the rocks can tell us much about its past history, both from the distribution and make-up of the strata themselves and from the fossils they contain. The whole of the Earth's surface has been mapped, sophisticated devices have been sent to chart the ocean deeps, and both instruments and humans have studied the Earth's atmosphere. Geophysical and geochemical data have provided answers to other problems, and together with the traditional techniques of geology, have got us to the position of realizing that our world has a segmented outer skin, the components of which move around slowly but inexorably as time passes.

Something else which we can now do, but which was inconceivable at the start of the twentieth century, is to look at our world from space, and from the surface of the Moon. It is these views, perhaps more than anything else, that have made us realize that the Earth is a planet, unexceptional in the Solar System except that it alone is suited to our kind of life.

The Earth in the Solar System

A casual glance at a plan of the Solar System shows how it is divided into two parts. First come four rela- tively small, rocky planets: Mercury, Venus, Earth and Mars. Beyond the orbit of Mars there is a wide gap in which move thousands of small rocky bodies called asteroids or minor planets. These are followed in turn by four giants worlds: Jupiter, Saturn, Uranus and Neptune, which are predominantly gaeous and quite unlike the Earth and its neighbours. Finally there is Pluto and its companion world, Charon, a curious double-planet system about which we know relatively little. Their revolution periods range from 88 days for Mercury to 248 years for Pluto/Charon: although

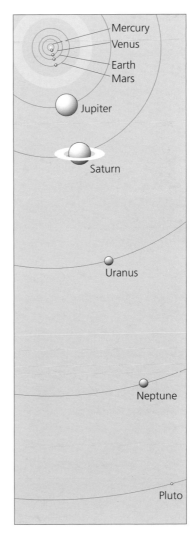

Orbits of the planets to scale. Note the clustering of the inner rocky group close to the Sun.

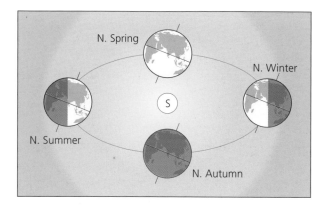

when near its perihelion (the point at which it lies closest to the Sun), the latter's eccentric orbit brings it within that of its giant neighbour, Neptune.

Of the inner planets, Mercury is the smallest, with a diameter of 4840 km. It has a cratered surface, superficially very like that of the Moon, and has virtually no atmosphere. For its size it is the densest planet of all, a fact that suggests something catastrophic affected it during its early history, probably removing a significant volume of the relatively less dense material which originally surrounded its metallic core. Venus is very nearly as large as the Earth, has a diameter of 12 100 km and an almost circular orbit located at a distance of 108 000 km from the Sun. Yet Venus and the Earth are not in the least similar. Venus has a dense atmosphere made up chiefly of carbon dioxide; a surface pressure 90 times that of the Earth, and a temperature which reaches over 500 °C. The clouds contain large amounts of sulphuric acid, making the planet little different from the conventional idea of Hell! Mars, at a mean distance of 227 000 km from the Sun, has a diameter of 6760 km and a revolution period of 687 days. Much smaller than the Earth, Mars has a tenuous, carbon dioxide atmosphere and permanent polar ice caps. The inclination of Mars's axis, being much the same as Earth's, means that it has seasons like ours, albeit far longer. However, the atmospheric pressure and temperature are far too low to support oceans, and current evidence suggests life does not exist there at present. The surface is partly cratered, is incised by ancient river courses and has immense (extinct) volcanoes which are far larger than any of their terrestrial counterparts.

The zone around the Sun in which temperatures are suitable for the existence of life, provided conditions are ripe, is often termed the ecosphere. Venus lies at the inner edge, and Mars at the outer, while planet Earth moves in the centre of the zone. It is likely that in the youthful Solar System, when the Sun was 30 per cent less luminous than it is now, Venus and the Earth started to evolve along similar lines; but when the Sun's luminosity increased the Venusian oceans boiled off and carbonates were driven out of its rocks, turning the planet into an inferno.

Although the giant planets are so massive (the mass of Jupiter is over 300 times that of the Earth) they have no effect upon our world simply because they are so distant. However, there are some asteroids that move away from the main swarm and approach the Earth; in 1937 a tiny asteroid, Hermes, passed by at less than twice the distance of the Moon. Occasional collisions can be expected and have no doubt occurred in the past. Indeed it has been suggested that an asteroid impact 65 million years ago led to a climatic change so violent that it caused the extinction of many forms of life, including the hitherto unrivalled dinosaurs. Such impacts may have caused or contributed to earlier mass extinctions too.

While there is little chance of an immediate massive impact, the Earth will not last for ever. Eventually the Sun will change its structure, and for a time will send out 100 times as much radiation as it now does. This will certainly mean the end of life on Earth, if not of the planet itself. However, the change in the Sun will not occur for at least 4000 million years, and probably longer, therefore there is no immediate cause for alarm!

Earth's orbit is not circular; at perihelion it lies 147 000 000 km from the Sun, and at aphelion 153 000 000 km. The revolution period is, of course, one year – more precisely 365.25 days; it is the extra quarter of a day that is responsible for the complications in our calendar. To prevent the civil calendar from getting out of step with the astronomical seasons, an extra day is added every four years; in these Leap Years February has 29 days instead of its usual 28. A further refinement in our present calendar is to omit 'century years' as Leap Years unless exactly divisible by 400. Thus 1900 was not one, but the year 2000 will be.

The seasons have little to do with the changing distance from the Sun. Perihelion actually occurs in December, when it is northern winter. The Earth's axis is inclined to the perpendicular to the orbital plane by 23.5°; in northern summer the North Pole is tilted towards the Sun, while in northern winter it is the turn of the South Pole to receive maximum warmth from the Sun's rays. Theoretically, southern winters should be longer and colder than northern ones, and the summers should be shorter and hotter, since the Earth moves fastest when near perihelion; but the difference is more-or-less cancelled out by the much greater expanse of continent-free ocean in the southern hemisphere. This tends to stabilize the temperature.

Over long periods the inclination of the Earth's rotational axis changes; this is due to the combined effects of the Sun and Moon which, because the Earth is not a perfect sphere, cause the axis to wobble rather like a spinning top. The axis describes a cone in the sky once every 26 000 years, a process called precession. Neither is the length of the day constant. Tidal friction between the Earth and Moon causes a slight but definite lengthening, easily measurable with modern atomic clocks. The orbital eccentricity also varies slightly, but this does not mean the Earth's orbit is unstable. There is no overall approach or recession from the Sun, which is fortunate, since even a slight variation would have a very marked effect on our climate.

The Earth's atmosphere and oceans

The Earth can broadly be called a 'wet' planet, as it has both an atmosphere and oceans of water; the latter make up the hydrosphere. Water is also locked up in the minerals and pore spaces within its rocks. It is unique in the Solar System in one respect: it is the only planet to have developed an atmosphere which we can breathe. Although the Earth intercepts a mere one part in 2 billion of the energy radiated by the Sun, and of that, a third is radiated back into space by the atmosphere, clouds and the surface, it is still this energy which drives processes at the Earth's surface. About a half of the radiation which does arrive is absorbed by the land surface and the oceans. Energy from the Sun also interacts with the atmosphere, and is the driving force for climate and the Earth's water cycle.

The atmosphere is composed largely of nitrogen (N_2: 78.08 per cent), and oxygen (O_2: 20.95 per cent), together with 0.93 per cent argon (Ar), and much smaller amounts of gases such as carbon dioxide (CO_2), neon (Ne), helium (He) and hydrogen (H_2). The present atmosphere is not, however, the original one. Early in Solar System history the outward-streaming wind of particles from the Sun would have stripped away most if not all of the primordial, hydrogen-rich atmosphere. The present air has been degassed slowly from the Earth's interior and initially would have held much more carbon dioxide than it now does, and much less free oxygen. It was only when plant life developed, perhaps 2000 million years ago, that carbon dioxide was removed and oxygen produced by the process of photosynthesis.

The lower part of the atmosphere is known as the troposphere; it extends upwards to an average height of 12 km, and in it resides roughly 90 per cent of the mass of the atmosphere. It is in this zone that we find our normal clouds and weather. The temperature decreases with height, as every mountaineer and pilot knows. Above the troposphere comes the tropo-

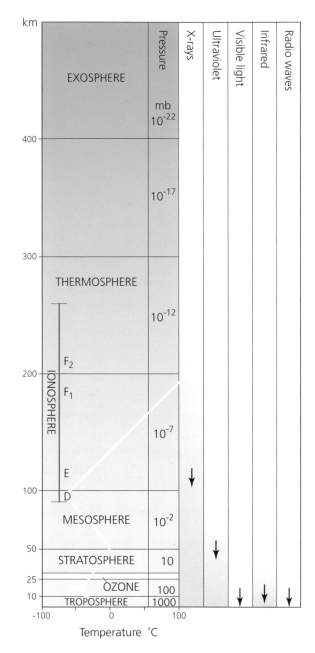

Structure of Earth's atmosphere. Variations in atmospheric pressure and temperature arise from the unequal distribution of solar heating. The troposphere is the zone of our weather and is separated from the stratosphere by the tropopause. It is within the stratosphere and the higher rarefied layers that most of the ultraviolet radiation is absorbed.

pause, which is in turn succeeded by the stratosphere, extending up to about 80 km. The temperature is more-or-less constant in this stratum, and it is here that we find the ozone (O_3) layer which is so important in the shielding of the surface from harmful short-wave radiations from space; without the ozone layer it is unlikely that life on Earth could have developed. The uppermost part of the stratosphere is termed the mesosphere.

Above comes the ionosphere, containing layers that reflect some radio waves and make long-wave radio communication possible. Finally there is the exosphere, which has no definite boundary, simply thinning outwards until its density is no greater than that of the interplanetary medium.

Part of the ionosphere is sometimes called the thermosphere and it is characterized by a very high temperature. This does not, however, mean that it is hot in the ordinary sense of the word. Temperature is determined by the velocities of the particles concerned; the faster the motion, the higher the temperature. In the thermosphere the velocities are high, and so is the temperature, but there are so few atoms and molecules at this height that the actual quantity of heat is inappreciable. There is a good analogy with a glowing poker and a sparkler firework. Each spark in the latter is white hot, but contains so little mass that there is no danger in holding the firework in the hand; however, it would be extremely unwise to grasp the poker, whose actual temperature is lower.

Aurorae are produced in the upper atmosphere. It is here that the Solar Wind – outwardly streaming, high-energy particles from the Sun – enters the Van Allen zones and 'overloads' them, so that charged particles cascade down toward the magnetic poles and produce beautiful auroral displays. Aurorae are most brilliant when the Sun is at its most active, which happens approximately every 11 years (this will next happen in the year 2002).

The oceans cover 70.8 per cent of the Earth's sur-

face, about 368 million km². The average depth is 3800 m and the total volume of water 1.4 billion km³. It is an enormous body of volatile material, and it is the evaporation of this, which transfers heat to the atmosphere, which is a vital factor in our climate. When water vapour condenses, heat is released and remains in the upper atmosphere. Thus the evaporation and condensation of water are responsible for transferring energy both vertically and horizontally from one place to another. This activity drives the water cycle, which plays a major rôle in influencing

Swirling clouds over the eastern shores of Newfoundland, USA, as seen by an Apollo spacecraft. [Photo: courtesy of NASA]

the appearance of the Earth's crust via rivers and weather.

The water oceans have the capacity to store enormous amounts of heat; indeed, the topmost 3 m of the oceans store as much energy as the entire atmosphere. Furthermore they can store or release large amounts of heat with only small changes in temperature. Heat is also transferred by ocean currents; cold currents flow from the polar regions toward the Equator, and warm currents in the opposite direction; they move in huge circular cells that are driven by the Earth's winds and rotation. Naturally, the motions are modified by the continents, the greater proportion of land in the northern hemisphere having a great effect on its climate and weather, and giving a very different pattern from that experienced in southern latitudes.

The Earth and the Moon

The differences between the Earth and Moon are largely a function of the Moon's much lower mass. The escape velocity is too low for it to have retained an atmosphere, and nor are there any traces of hydrated minerals. The Moon is a lifeless world and always has been. At one time it was believed that it had been a part of the Earth which had separated off; however, we now are fairly sure that soon after the Earth's core had formed, a massive planet-sized object gave the Earth a glancing blow. The object was vaporized, and the Moon formed largely from its mantle. This neatly explains why the Moon and Earth are chemically dissimilar and why the Moon has lost most, if not all of its volatiles.

The Moon has a diameter of 3475 km and a mass 1/81 that of the Earth. It has a mean distance from the Earth of 384 000 km and its revolution period is 27.3 days, although because the Earth–Moon system is in orbit around the Sun, the lunar synodic period (that is to say, the interval between successive new moons) is 29.5 days. It is the controlling force producing ocean tides. The Moon pulls upon the Earth, raising a bulge in the ocean surface; thus the water heaps up under the Moon at high tide, with a similar high tide on the opposite side of the Earth. As the Earth rotates, the bulges do not rotate with it, and so the tides sweep across the planet's water surface. In practice the whole situation is extremely complex, as we must also reckon with the tide-producing forces of the Sun, which are weaker than those of the Moon but nevertheless very evident. When the Moon and Sun are pulling in tandem (at new or full moon) we have strong or 'spring' tides; when the two are pulling against one another (at half-moon) we have weaker or 'neap' tides.

It is tidal friction which is slowly increasing the length of the day. In the past, before the Moon solidified, tides in it produced by the Earth slowed down the rotation so much that, relative to the Earth, it stopped rotating altogether; today the Moon keeps the same face turned toward the Earth at all times. It should be borne in mind, however, that the Moon does not keep the same face turned toward the Sun, so that day and night conditions are the same all over the lunar surface.

The message of asteroids and meteorites

Millions of small bodies exist in the Solar System. Among these, the asteroids (minor planets) form a prominent group. Most of these are confined to orbits that lie between those of Mars and Jupiter, although some do have Earth-crossing paths. The largest is called Ceres. This was discovered by the Italian astronomer Giuseppe Piazzi in 1801; it measures 1003 km across. The smallest yet found (6344P-L) is a mere 200 m in diameter. It is estimated that there are over one million asteroidal objects with diameters in excess of 1 km.

Asteroids originally resembled the rocky fragments that accreted to form planets. However, before they could stick together, the gravitational pull of the Sun and the planets sent them into tilted, elongated orbits. Collisions occurred and their fragmentation continued. This destruction continues – somewhat more slowly – until the present time. The powerful gravitational influence of Jupiter, as it grew, must have inhibited planetary growth in its vicinity, yet would have collected together bits and pieces in the belt where the asteroids now find themselves.

When seen through a telescope many asteroids exhibit changes in brightness; this may be a function of irregular shape, but also indicates differences in chemical composition between one and another. The most abundant ones are type C, or carbonaceous asteroids. These are darker than coal and tend to orbit in the outer regions of the Asteroid Belt. Silica-rich asteroids are termed S-type bodies, are of intermediate albedo, and orbit in the middle parts of the

Chondrite meteorite. [Photo: courtesy of NASA]

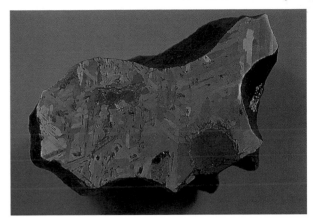

Iron–nickel meteorite. [Photo: courtesy of NASA]

Belt. M-class asteroids are metallic and probably represent the metal-rich cores of larger parents that disrupted long ago.

More abundant than asteroids are meteorites. They show a similar range in chemistry, are generally much smaller and have Earth-crossing orbits. It is for this reason that meteorites hit the Earth and can be collected for study. In essence they are free samples of Solar System rock. Very accurate tracking of some meteorites allows calculation of their orbits which seem very similar to Earth-crossing asteroids such as Apollo and Icarus-2. It appears that these may have once resided in the main asteroid belt but were knocked out of it by the strong gravitational influence of Jupiter.

Traditionally meteorites are classified as stones, irons and stony-irons – a similar division to that applied to asteroids. However, a more meaningful classification sees them either as 'differentiated' or 'undifferentiated' types. Within the latter class are the chondritic meteorites which contain chemical elements in similar proportions to the Sun's atmosphere. Radiometric dating of these samples reveals them to be very ancient, about 4.5 billion years (4.5 Ga). They represent some of the most primitive samples we have yet found within the Solar System.

Within the chondrites are high-temperature inclusions rich in aluminium (Al), also volatile-rich particles, and peculiar spherical objects called chondrules which are the result of melting. The presence of such diverse material (which is why they are considered 'undifferentiated') within such ancient meteorites appears to indicate that the stuff from which the planets eventually accreted was rather diverse but well mixed at an early stage in Solar System history.

The differentiated types show signs of having undergone changes; either melting or metamorphism, i.e. heating and recrystallization. They appear once to have been more primitive bodies which underwent modification during early Solar System history. Their ages are similar to those of their undifferentiated colleagues. Some types, however, are rather different and yield much younger ages. Most interesting of these is the small group termed the SNC group (named after three meteorite samples called Shergotty, Nakhla and Chassigny, respectively). Such samples have a gas content that is very similar to that of the Martian atmosphere, are much more oxidized than is usual, while their average age is only 1.3 billion years (1.3 Ga). Together these characteristics indicate that they come from a differentiated planetary body large enough to have undergone magmatic activity at that time. Mars seems the most likely candidate – all things considered – and one view is that an oblique impact caused them to spall off Mars and reach velocities of at least 5 km/s.

Is the Earth unique?

We have seen that in many ways the Earth is just a rather ordinary planet, yet it appears to be the only one where conditions have remained suitably benign for the development of life. Certainly it is unique in terms of the development of the human species within the Solar System. This begs the question of the Earth's status regarding uniqueness in the Universe.

The life question has always been fascinating and remains so until the present time. Interestingly, if we look back in time, to a point well before the human species had evolved, the dominant animal group clearly was the dinosaurs. These had evolved in such a way as to be able to inhabit a wide range of environments for a very lengthy period of time; some were frighteningly aggressive and undoubtedly ruled the roost! Yet, around 65 million years ago they disappeared from the record, almost at a stroke.

One suggestion that has been made is that a massive impact (craters of an appropriate age have been discovered in the Gulf of Mexico and Yucatan) had such an effect on the Earth that a drastic climatic change occurred, rendering conditions unsuitable for these large beasts. Their habitat was effectively taken away, food became too difficult to obtain, and they simply could not adapt and died out. This mass extinction appears to have made way for the human species. The interesting spin-off from this idea is that, since massive impacts can only occur by complete chance, had it not happened the human species may never have developed (the dinosaurs might still be the ruling class). In other words, we, the highest order of animal life, may not have been the ultimate result of evolution had things been different; or would nature have found some other way to achieve this inevitable goal?

At present it seems we have no answer to this question. However, the recent reporting of the discovery of tiny bacteria-like objects in a SNC-like meteorite collected in Antarctica, has been interpreted by some scientists to point to the strong possibility that some form of life may have developed on our neighbour planet, Mars. It has also been revealed that water oceans very likely exist beneath the icy crust of Jupiter's large moon, Europa; oceans in which conditions may have once provided suitable conditions for some form of organic growth. Again, while Venus has long been an extremely hostile world, in the very distant past it also may have had warm oceans; who is to say whether or not life once existed there too? Perhaps Earth cannot lay claim to be the only life-bearing world in our Solar System after all!

Translating this to the wider perspective of the Universe at large: there is no logical argument that can promote the Earth to the status of a unique life-bearing planet. Somewhere in space must be thousands of suns which have planetary systems, or discs of dusty material from which such could develop; surely some of these must have seen development of some kind of intelligent life? Certainly the search for other 'accretion discs' and planet-sized companions to Sun-like stars is high profile just now. The A-type star of ß Pictoris was the first probable proto-planetary system to be detected, it lies 78 light years distant. The astronomical observations of this object suggest that it is surrounded by a disc of material that extends 80 billion km from the central star which itself is 68 times as luminous as our own Sun. One day a planetary system will doubtless evolve out of this immense cloud of matter.

ß Pictoris is a relatively near object (which is why it was discovered in the first place); similar stars and Sun-like stars with accretion discs or planetary systems must surely exist in other galaxies, well beyond the reach of any of our current instrumentation. The immensity of the universe ensures anonymity for any life which lies out there. Earth thus appears unique but appearances could be very deceptive.

2

The formation of the Earth

Creation or Big Bang?

What can we really say about the Creation? Probably no subject has caused more discussion – and more dissent – over the years, and even today we cannot claim to know much about the beginning of the Universe itself. But insofar as the Earth is concerned, we have at least some facts to guide us, and in particular we are reasonably confident about the time scale. The Earth – and the other planets – began their separate existence about 4700 million years ago (4.7 Ga); the Universe is older still, and one reasonable estimate leads us to a figure of 15 billion years (15 Ga). According to the current school of thought, all matter came into existence suddenly, in what is termed the Big Bang. Once created, it began to expand; in the course of time galaxies were formed, then stars, and then planets.

What we do not know is the reason for the initial Big Bang. The matter must have come from somewhere so, if it originated in a huge explosion, what happened before that? The only plausible alternative to this idea is to assume that the universe has always existed, and will never die. In the late 1940s the 'Continuous Creation' was proposed by a group of scientists at Cambridge University, and was later refined and elaborated upon by Sir Fred Hoyle; the expansion of the galaxies was accepted, but it was claimed that old galaxies, when they passed beyond observable range, would be replaced by new ones, created from material which appeared spontaneously in the form of hydrogen atoms. This so-called Steady State Theory was quite popular for a time, but gradually observational evidence against it mounted up, and today it has been almost totally abandoned. The only logical alternative to the Big Bang appears to be the 'oscillating universe' idea, which makes the assumption that the present phase of expansion will be followed by one of contraction, until all the matter reassembles, generating a new Big Bang.

Going back to the seventeenth century, we find

that James Ussher, Anglican Archbishop of Armagh, solved the whole problem neatly by adding up the ages of the Old Testament patriarchs and making similar calculations which satisfied him and many others. In 1650 he stated categorically that the world had begun its career at ten o'clock in the morning of 26 October 4004 BC. Subsequently it was shown beyond doubt that many fossils are much older than that, and a more scientific explanation was sought.

Early theories

One interesting attempt at explaining the Earth's origin was made by a Frenchman, the Comte de Buffon, in 1779. He argued that the Earth must be at least 75 000 years old, and could have been produced by a collision between the Sun and a comet, with the result that the material was sprayed about in sufficient quantities to form planets. While such an idea is very wide of the mark, since comets are very flimsy objects that could not possibly cause such a catastrophe, at least he extended the time scale a little. Furthermore, his contemporary, Mikhail Lomonosov, Russia's first great astronomer, was considering an age of a few thousand years. Different ideas were put forward by the philosophers Immanuel Kant in Germany and Thomas Wright in England, who believed that the planets were formed from a cloud of material associated with a youthful Sun. This type of 'Nebular Hypothesis' was refined and improved in 1796 by Pierre Simon, Marquis de Laplace, who was an excellent mathematician as well as a far-sighted scientist.

According to Laplace, the Solar System began as a vast cloud of gas, disc-shaped and in slow rotation. Gravity caused it to contract; as it did so, its rate of rotation increased, until the centrifugal force at its edge became equal to the gravitational pull there. At this stage a ring of matter broke away from the main mass, and collected into a planet. As the contraction continued, a second ring was thrown off, and so on. The planets were conceived as having formed one-by-one, and the remnant of the original cloud made up the present-day Sun.

It all looked quite neat and tidy, and Laplace's Nebular Hypothesis remained unchallenged for many years, but gradually some rather awkward facts emerged. Mathematical analysis showed that rings would not be thrown off during contraction of a cloud; and even if this were possible, these rings would not condense into planets. Worse still was the problem of angular momentum, a quantity derived by considering one body moving around another and combining its mass, distance and velocity (angular momentum = mvR, where m is the mass of a particle, moving in a cicular orbit of radius, R, at a velocity, v). It is a fundamental principle that while angular momentum can be transferred, it can never be destroyed. By Laplace's theory all the angular momentum possessed by the Sun and planets must originally have resided in the gas cloud, and most of it would now be concentrated in the Sun. This meant that the Sun should be rotating fairly quickly. Actually, the Sun's rotation period is 25 days, and almost all of the angular momentum of the Solar System resides in the four planets Jupiter, Saturn, Uranus and Neptune. There seemed no way of overcoming this difficulty, and the Nebular Hypothesis was reluctantly discarded, to be replaced by a crop of 'catastrophic' theories involving the action of a passing star.

The first of these was propounded by two Americans, the geologist Thomas Chamberlin and the astronomer Forest Ray Moulton, in 1900. Space is sparsely populated. The Sun, which is a normal star, has a diameter of 1 392 000 km. However, if we were to represent it by a tennis ball its nearest stellar neighbour would reside over 1000 km away. Therefore, collisions or even close encounters of two stars must be very rare, at least in our part of the Universe.

However, we cannot entirely rule out an encounter, and the two Americans believed that as a wandering star approached the Sun, a cigar-shaped tongue of matter was torn away from the solar surface. As the intruder receded, the tongue of material was left spinning around the Sun; slowly it condensed into planets, with the largest bodies, Jupiter and Saturn, in the middle of the Solar System, where the thickest part of the 'cigar' would have been.

The passing-star theory was supported and popularized by Sir James Jeans, remembered today not only for his contributions to theoretical astrophysics but also for his books and broadcasts. It, too, looked remarkably neat; it would mean that planetary systems would be very common in the Galaxy. Yet again there were mathematical problems, and attempts at further refinement were no more successful. The best of them came from Sir Harold Jeffreys, who suggested that the passing star struck the Sun a glancing blow.

The next theories involved a binary companion of the Sun. Star pairs or binaries are very common, and there was nothing outrageous in suggesting that the Sun also might once have been one component of a binary system. The American astronomer Henry Norris Russell believed that it was the companion star that was struck by the intruder, causing enough debris to form the planets; Raymond Lyttelton considered that a near approach from a wandering star would wrench the binary companion away from the Sun, planet-forming material being scattered in the process; Sir Fred Hoyle proposed that the solar companion exploded as a supernova, shedding much of its material in the Sun's neighbourhood before departing permanently.

An interesting modification was due to Gerard Kuiper, who regarded the Solar System as a 'degenerate binary', in which the second mass did not condense into a star, but remained as dispersed material. Therefore, the end product was one star (the Sun) and several condensations of proto-planets, which gradually formed into the planets of today. Once a proto-planet became sufficiently massive, it would draw in material as its gravitational pull increased.

Modern hypotheses

Recent ideas are different. Passing and binary stars are generally not in favour, but mention should be made of the theory proposed by Michael Woolfson of York University, who suggests that the Sun was produced as a member of a whole cluster of stars, relatively close together. If the Sun's formation were completed at a fairly early stage in the history of the cluster, it could have been approached by a 'proto-star' at an earlier stage in its evolution; the Sun would then have pulled a long filament of material away from the proto-star, and this filament could have broken up into globules from which the planets subsequently formed. Woolfson's hypothesis is no means unreasonable, but in general most modern theories are much more like Laplace's Nebular Hypothesis than those of the catastrophists.

The trend in this direction was set in the 1950s by Otto Schmidt in Russia and Carl von Weizsäcker in Germany, whose ideas were similar in principle though different in detail. They assumed that the youthful Sun was surrounded by a 'Solar Nebula', a vast envelope of material made up largely of the two lightest elements, hydrogen and helium, together with what loosely may be termed dust. Collision and friction between the Sun's family of particles led to formation of a disc-shaped cloud, from which the planets were built up by the less than perfectly understood process of 'accretion'.

Basically, this kind of idea is still supported today, and it is usually considered that the Solar System began due to contraction of the Solar Nebula. Rotation meant that the nebula would have become disc-shaped and that planets would then form by accretion, a poorly-understood method by which col-

Diagram to show the condensation of the planets from a rapidly-rotating Solar Nebula.

lisions and rebounds both abrade and add mass to the particles within the cloud. However, this did not occur in the manner envisaged by Laplace. The outermost regions of the nebula were naturally the coolest, and substances such as water, ammonia, methane and other volatiles condensed; the embryonic outer planets consisted mainly of these materials which, when they had become sufficiently massive, were able to draw in hydrogen and helium in huge amounts. They ended up as the four giant planets with which we are familiar today. Closer to the Sun things were rather different. Here, the temperature was higher, and planetary growth followed a different pattern. While the temperature prevailing in the Solar Nebula remained generally high (around 1000 °C) dense, iron-rich compounds precipitated in the inner regions, forming the inner planet cores. When temperatures had fallen sufficiently, silicate

materials of lower density could be swept up by the growing planetesimals. This explains why the Solar System is divided into two parts. The fact that there is no major planet between Mars and Jupiter can be accounted for by the powerful gravitational field of Jupiter, which disrupted any embyonic planet in this region before it had a chance to grow, producing instead the swarm of dwarf worlds which are the asteroids.

If this is correct, we may expect all the planets to be of roughly the same age, and it is certainly true that the Moon, which is probably regarded as a companion planet of the Earth rather than as a mere satellite. Not everyone agrees with this theory, however, for instance A. J. R. Prentice of Monash University in the USA has reverted to a modified form of the old Nebular Hypothesis, which he believes is a real possibility, provided that the Solar Nebula collapsed quickly enough: in which case the giant planets would be older than the Earth. The Swedish physicist Hannes Alfvén has emphasized the importance of magnetic fields, and believes that in its early years the Sun was rapidly rotating but without planets; later, its strong magnetic field resulted in the sucking-in of material, which condensed into planets and at the same time retarded the Sun's rotation.

Quite clearly there are still marked divergences of opinion, but two things do seem more or less established: the Earth and other planets were formed between 4500 and 5000 million years ago from a Solar Nebula, and the smaller inner planets were relatively hot from an early stage. It also seems certain that in the early stages of Solar System history the Sun was less luminous than it now is, perhaps by a factor of one quarter. It was also somewhat unstable, and ejected material in the form of a strong Solar Wind. This, too, is supported by observational evidence, since we know of very young stars, known as T Tauri stars, which are in this stage now. The T Tauri wind would have blown much of the thinly spread material away from the inner region of the Solar System, and would also have taken most of the angular momentum away from the central Sun.

The origin of the Moon

Sir George Darwin (son of Charles Darwin) suggested in 1878 that the Earth and Moon used to be a single body which was in rapid rotation; the whole mass became pear-shaped and then dumbbell-shaped until the 'neck' of the dumbbell broke and the Moon moved away independently. Later it was even suggested that the great depression of the Pacific Ocean marked the place from where the Moon departed. Both are highly improbable – for instance, the Pacific is a relatively shallow basin, having far too little volume to account for the birth of a body one quarter the size of the Earth, while the Moon is chemically too dissimilar to the Earth to once have been a part of it.

The mean density of the Moon is significantly less than that of the Earth and, since much of the Earth's high average density derives from the heavy iron/nickel-rich material in its core, this suggests that the Moon does not have a large dense core. The chemical make-ups of the Moon and Earth also show certain critical differences: for a start, the lunar rocks lack hydrated minerals, which are commonplace within the Earth. There are also differences in the amounts and proportions of certain trace elements which, if the two had a common origin, should bear a far closer resemblance. There are other differences too. Modern research suggests that soon after the Earth's core had formed – probably within a few hundred million years of the Earth's birth – a massive planetesimal gave the youthful Earth a glancing blow. The object was vaporized but the Moon was then formed largely from the object's mantle. This explains why the two worlds are chemically so dissimilar and why the Moon lost most of its volatiles early on.

Exactly how old is the Moon? Samples of lunar rocks have been accurately dated, using radiometric techniques. The age of the Moon, derived by these means, comes out as 4.6 billion years (4.6 Ga). The oldest actual samples – of the lunar highlands – yield ages of around 4420 million years. It is also clear is that a major chemical differentiation had taken place within the Moon by 4400 million years ago because, by this time, the reservoir of magma from which the mare basalts eventually rose towards the surface, had been separated off from the highland material. The mare basalts – which infilled ancient impact basins within the lunar highland crust – yield ages which are less than this, many being younger than 3800 million years.

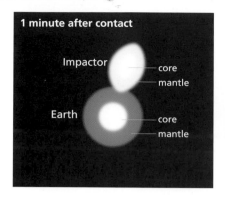

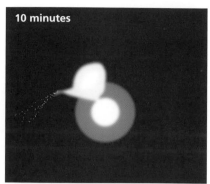

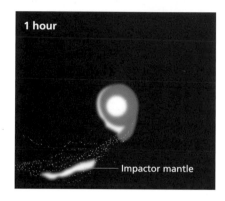

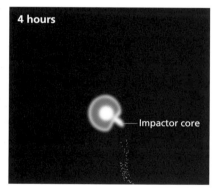

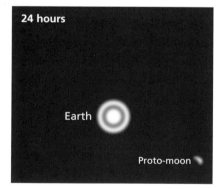

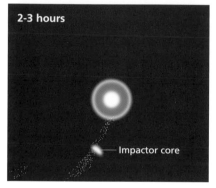

Diagram showing the origin of the Moon by a grazing collision with a large planetesimal.

The age of the Earth

The Moon has remained inert for the greater part of its long history which is in sharp contrast to the Earth. The latter has remained dynamic until the present day, recycling its crust, generating earthquakes and volcanic materials. This means that any potential remnants of the primordial terrestrial crust will have long since been recycled inside the planet. The oldest dated rocks have yielded ages of between 3900 and 4100 million years; while this is somewhat younger than the oldest lunar samples, this does not mean that the Earth is younger (indeed, this cannot be so).

It is generally assumed that the Earth, like the Moon, was formed at the same time as the other planets and that the ages yielded by meteorite samples, i.e. 4600 million years, lies close to this age. We live on an ancient world which has changed constantly and significantly since its formation; however, the changes are not so complete that tracts of very ancient rocks have not been left for geologists to study. There are large areas of very old rocks which help us in building up a picture of how our planet has changed through the aeons of time.

3

The primaeval Earth

There may be heaven; there must be hell;
Meantime, there is our Earth here – well!
ROBERT BROWNING *'Time's Revenges'*

Birth of a planet

Our conception of the Earth's birth has changed
many times. Even during the 'space age', which began
approximately when the Russians launched their first
successful Sputnik in 1957, our ideas on this subject
have been in a state of flux. Today there is no room for
a reasoned belief that the Earth was suddenly con-
jured up from some universal 'stuff', at a specific
moment in time. Our knowledge of the chemical and
physical processes that operate within the Universe
has taken leaps forward since the days of, say,
Newton. Even so, we must exercise considerable cau-
tion, as there are many unknowns, even in the most
modern hypotheses. In a way, the evolution of stars is
still better understood than that of planetary systems.

Intensive studies of meteorites have led us somewhat nearer to what reasonably may be considered
the truth, while studies of our neighbour planets,
asteroids and comets have added to our understanding considerably. We are now fortunate to have samples of various types of Moon rock and to know their
ages; these have been analysed in terrestrial laborato-
ries. Spacecraft have also landed on Venus and analysed surface rocks *in situ*, while the Viking Land
Pathfinder lander probes sampled and analysed the
regolith in three different regions of Mars. All of this
information, when coupled to what we have learned
of the processes that have moulded and modified the
surfaces of our nearby neighbours in space, has enabled us at least to outline the time scale and sequence
of events that were involved.

The modern view is that the Earth, like the other
Solar System bodies, accumulated rather quickly
from a cloud of dust and gas surrounding the proto-
Sun, i.e. the Solar Nebula. That the process was rapid
is indicated by the presence of decay products from
several short-lived radioactive isotopes in certain
meteorites, e.g. ^{26}Al (half-life 10^6 years), ^{29}I (half-life
16×10^6 years) and ^{244}Pu (half-life 82×10^6 years).

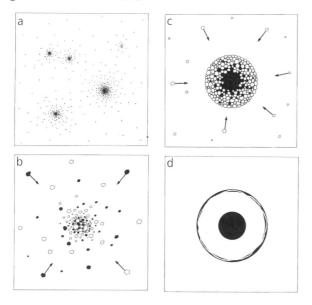

Accretion of Earth: (a) dust particles form in the nebula; (b) larger particles are accreted onto one another; (c) a planet-sized body forms with heavy elements sinking to form a dense core; (d) the outer mantle builds up.

(Some atoms have radioactive isotopes which decay with time, at a known and fixed rate. Their decay products are non-radioactive. The time taken for exactly one half of the original amount of a radioactive parent isotope to decay to its non-radioactive daughter isotope, is termed the half-life. By measuring the relative proportions of radioactive and non-radioactive nuclides, it is possible to determine the age of a sample.) The fact that these rapidly-decaying isotopes have left traces at all, implies that the particles containing them must have been very rapidly incorporated inside the body in which the meteorites originated.

Of further interest is the observation that such isotopes are not found in all meteorites, only some. Now if the parent radionuclides were indigenous to the Solar Nebula, their residues ought to be equally distributed throughout all meteorites. Scientists believe the most plausible explanation for such an unequal distribution is that the elements had been expelled from a nearby supernova explosion, and were captured by the Solar Nebula, to be implanted in some chondritic meteorites.

The quickly-growing meteorites have also retained important chemical and physical evidence as to what went on during the formative stages of planetary growth. The group known as the chondrites are of particular importance. These bodies, which have a radiometric age of around 4600 million years – making them the oldest dated samples of Solar System

material – are taken to represent very primitive material whose age must be close to that of the system itself. Surprisingly they contain fragments, not only of very-high-temperature materials, but also of volatile material, and matter which shows clear signs of having melted and chemically differentiated. Clearly, therefore, the body or bodies from which these fragments came must have been made up from each kind of particle which existed side-by-side in the Solar Nebula (planetesimals)

Little factual evidence exists concerning how accretion actually occurred. It is widely believed, however, that at some point nuclei grew within the nebula, and that from these tiny aggregations of matter the planets gradually grew. In the case of the Earth, it is thought that growth began with dust-sized grains which, with the assistance of weak electrostatic forces, were converted into centimetre-sized particles. In time these aggregated into larger and larger bodies until first asteroidal and then planet-sized bodies evolved. The process must have been highly complex; probably periods of growth alternating with periods of disruption and then reaggregation, this occurring many, many times before growth was completed.

All of the larger solid bodies within the Solar System were subjected to intensive bombardment by planetesimals and meteorites during the first 500 million years of Solar System history. The surfaces of all of the rocky inner planets, and of most outer planet moons and asteroids, are heavily cratered, each crater being the scar left by a single impact event. It is evident that it was the repeated collisions between bodies of differing sizes which eventually led to planetary growth. The longer the surface of a body has been exposed to impacts, the more craters it seems to have collected, assuming no other process obliterated the craters formed. This fact is the basis of one important method of establishing the relative ages of different surfaces.

The kinetic energy of the impact process is ultimately converted into heat; thus, as a primitive planet grows, so it attracts larger and larger planetesimals, these impacting at higher and higher velocities. Eventually substantial amounts of kinetic energy are implanted inside growing planetesimals as thermal energy; for this reason the surface layers would eventually become much hotter than their interiors. Much of the heat would, of course, be radiated away into space, but where very massive impacts occurred, some thermal energy would have been implanted deep inside a planet of Earth size; this would not escape easily.

Both Earth and Venus have substantial cores of dense metallic elements that must be at very high temperatures indeed. The production of such cores must itself have been a very important heat-generating process, as the downward movement of droplets of dense core-forming elements produces gravitational potential energy that would raise the internal temperature even higher. Gradually, therefore, the Earth would have become hot and, from what we have learned about the Moon in recent years, large parts would actually have melted, possibly producing what is termed a 'magma ocean' – huge volumes of molten silicate material – close to the surface. These may well have formed the outer layers of Earth-type planets. Gradually they would have cooled, whereupon there would have been a slow thickening of the solid outer layer which we know as the crust or lithosphere.

The primary atmosphere
As the proto-planets began to evolve, so too did the proto-Sun. The inward pressure of gases within the Solar Nebula eventually reached such high levels that thermonuclear reactions were triggered. At this time the Sun was truly 'born'. Many young stars have been studied, and it has been shown that at this stage in a

star's history a large amount of gas is blasted away into space. This is achieved by the activity of what is termed the 'Solar Wind', whose importance was first discovered when it was studied from spacecraft during the 1960s.

Essentially, the Solar Wind consists of fast-moving protons and electrons which, by the time they reach the Earth's orbit, are travelling at velocities of about 400 km/s. Although the Solar Wind is very tenuous, with a particle density of only a few tens of particles per cubic centimetre, its effects are far-reaching. In the case of a newborn star such as the proto-Sun, the wind becomes a veritable gale; this is during what is termed the T Tauri stage (the name comes from the

The outward-moving Solar Wind stripped away the primordial atmospheres of the terrestrial planets at an early stage.

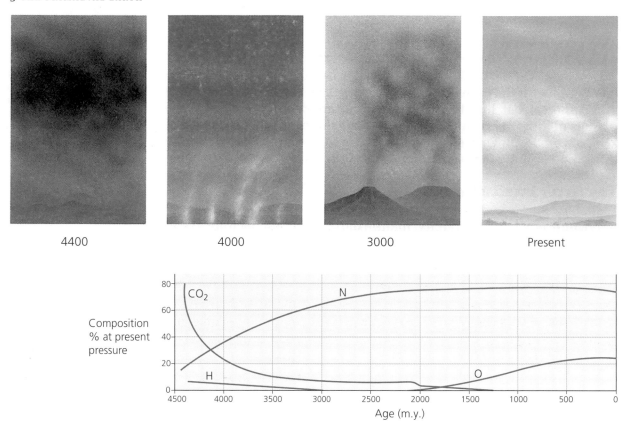

4400   4000   3000   Present

Schematic diagram of how Earth's atmosphere evolved. The decreasing amount of carbon dioxide was balanced by an increase in the nitrogen content. With the rise of plant life, about 2500 million years ago, free oxygen began to enter the atmosphere.

first star discovered to be at this point in its evolution).

The Earth's primordial atmosphere was almost certainly rich in water vapour, carbon dioxide, carbon monoxide (CO), nitrogen, hydrogen chloride and hydrogen. Most of the hydrogen quickly escaped into space, while some of the water vapour in the upper atmosphere was broken down by sunlight into hydrogen and oxygen, the latter escaping and combining with gases like methane (CH_4) and carbon monoxide to form water (H_2O) and carbon dioxide. The remaining hydrogen and helium was stripped off during the Sun's T Tauri stage, and blasted out of the Solar System into interstellar space. This cosmic cleaning up process saw the stripping away of the primary atmospheres of all of the inner planets, so that only their solid parts remained. Therefore only the planets, asteroids and meteoroids survived this first stage, and most of the mass of the Solar System (99.8 per cent) has remained in the Sun.

The Earth's present atmosphere is entirely secondary, and has been derived from the interior by a process of degassing, which occurred as the mantle layer solidifed. By the time that this took place, much of the metallic iron had sunk towards the core, and oxidizing conditions promoted by the relatively oxidized mantle resulted in the production of oxidized gaseous forms of nitrogen, sulphur and carbon (i.e. NO_2, SO_2 and CO_2). Confirmation of this secondary origin is to be found in the relatively low abundances of the noble gases neon, krypton, and xenon on the Earth as compared with their abundances in the Sun.

Further modification of Earth's atmosphere has taken place because of biological processes such as photosynthesis. Over the last 3000 million years, living organisms have relentlessly changed the atmosphere in such a way that our present air is quite different from the atmospheres of Venus or Mars, planets that in many ways are not too unlike the Earth. The Earth differs also in being largely covered with water; over 70 per cent of our globe surface is ocean.

Research indicates that, during their early lives,

the inner planet atmospheres were probably little different from each other. The subsequent changes occurred due to their different locations within the Solar System. However, Venus, Earth and Mars can be treated as a group when compared with the outer planets; furthermore whereas the terrestrial planets have atmospheres that hug their surfaces and are relatively shallow, the planets of the outer group have gaseous envelopes tens of thousands of kilometres deep. This striking difference is the direct result of the differing positions of the planets and of the composition of the primordial grains from which the various planets accreted.

The Great Bombardment

Planetary surfaces are moulded by two types of process: internal (e.g. volcanism, mantle motion, tectonics) and external (e.g. meteorite impact, weathering and erosion). Traditionally geology has not included impact studies but, with the onset of planetary exploration by spacecraft, it quickly became apparent that the rôle of impact processes was of extreme importance. This realization first came from studies of the Moon, whose surface shows the scars of impacts and the deposits of related processes.

Today the Solar System comprises a relatively ordered collection of larger objects – the planets, their satellites and the Sun; and a group of smaller, somewhat less orderly bodies – the asteroids, comets and meteoroids. In very early times the proportion of disordered to ordered material was far greater than now, and collisions between proto-planets and this loose debris were commonplace. Since a collision between a planet and a fast-moving fragment is a very energetic event, considerable damage can be done to both target and impactor.

Spacecraft images of Mercury, the Moon and Mars and, indeed, many of the moons of the outer planets, show their surfaces to be pockmarked with craters of varying size, as well as some larger impact basins. Most of these were produced by collisions with wandering planetesimals during a period of intense bombardment which took place over 4000 million years ago. This phase of intense geological activity has been termed the Great Bombardment.

The amount of energy released during a fairly typical impact can be appreciated from the following example: consider a 10-m-diameter iron-rich meteoroid with a relative velocity of 10 km/s. The kinetic energy of such a body impacting the Earth is 1.36×10^{22} erg/g. The bare figures may not be particularly meaningful, but when we consider that one kiloton of TNT has a chemical energy of roughly 4.2×10^{10} erg/g, its potential for destruction becomes readily apparent. In fact the figure quoted is equivalent to a moderate-strength earthquake! Unless such a body was so decelerated or abraded during its descent that it lost virtually all its mass (which would occur only if the body were very small), its impact on the Earth's surface would blast out a crater with a volume thousands of times that of the original impactor. The impactor would disintegrate, and would be dispersed along with the terrestrial debris thrown out of the cavity.

Thus far very little evidence has been collected which shows that impacts have left their marks in very ancient terrestrial rocks; however, this does not imply that impact events did not occur. On the contrary, they must have done so; calculations based on scaling from lunar experience, suggest that more than 100 impact structures with diameters in excess of 1000 km should have formed on Earth. Very recently the discovery has been reported from both South Africa and Western Australia, in Archaean rocks with ages of between 3.2 and 3.5 Ga, of bands of rock 30 cm to 2 m thick, containing sand-sized silicate spherules that may be the first direct evidence for impact melt droplets. The beds extend over at least 100 km laterally, and appear to be associated

Wolfe Creek impact crater, Western Australia. The crater is just under 1 km across and was formed less than 0.3 million years ago.

Map showing the distribution of the Earth's suspected impact craters.

with shock features. So who knows what may be unearthed as research proceeds with a new awareness!

About 100 impact craters have been partially preserved, including famous structures such as the Barringer Crater in Arizona and Wolfe Creek in Western Australia. Some are better preserved than others, the Barringer Crater – with an age of around 0.35 million years – being particularly striking in its state of preservation. The great majority of craters are found on the surfaces of the Earth's ancient shield areas, such as Australia, Canada and Africa. These have remained stable for very long periods, have not suffered translational movements and suffered only relatively modest erosion. The rarity of such craters when compared with their numbers on, say, the Moon and Mercury, is due in part to the comparatively rapid rate of weathering and erosion on the Earth, and in part to the restless nature of the Earth's lithosphere, which is constantly recycling itself over relatively short periods in geological terms.

Most geologists would subscribe to the view that the Great Bombardment really did affect the Earth, but some experts in the field of lithospheric development – the speciality which deals with the properties and development of Earth's crust and lithosphere – will probably argue that the Earth was too mobile for it to have had a primitive surface akin to that of the Moon. What is perhaps most interesting is that the earliest life had appeared by about 3500 million years ago, perhaps earlier, just after the period of most intense bombardment had ended. Did the close of this phase make way for a very different Earth? Well, there certainly is strong evidence to suggest that the volatiles necessary for the atmosphere and hydrosphere were present at this time, and that these formed the basis for organic development. Some scientists have argued that these volatiles were brought to the Earth by comets and meteoroids. If this were so, then it would clearly be true to say that the close

of the Great Bombardment marked the beginning of a new epoch in Earth history.

The primitive Earth

The way in which the individual planets evolved and, on the broader scale, why the Sun's family of planets developed into two quite distinct groups of worlds, was largely a function of their distance from the Sun. In its infancy the Solar Nebula was probably rather well mixed, with a temperature approaching 2000 °C, and was mainly gaseous. Temperature had a profound effect on exactly what happened next inside the Solar Nebula, because various kinds of matter have rather different chemical characteristics. Thus Mercury, Venus, Earth and Mars accreted close to the proto-Sun where nebular temperatures were high; the outer planets – Jupiter, Saturn, Uranus and Neptune – on the other hand, grew where nebular temperatures were at or below 0 °C and where ices and gases condensed out of the primordial cloud.

Temperatures were highest close to the proto-Sun, the first solid particles to condense being refractory materials composed of metal oxides, chiefly of tungsten, aluminium and calcium. As the nebular temperature slowly fell these reacted with gases in the cloud to form silicates, chiefly of magnesium, iron and calcium. The inner planets and asteroids are largely made from such materials.

Further from the Sun, where temperatures typically hovered around the 50 °C mark, compounds rich in carbon would have condensed. Water ice may have existed as snowflakes and become incorporated into the growing grains. Further out again, volatile elements such as argon and compounds like ammonia and methane would have been the norm, i.e. the ices and liquids were combinations of carbon, nitrogen, hydrogen and oxygen (e.g. water, H_2O). Finally there were those substances which, except under abnormal conditions, are gaseous, e.g. hydrogen,

Rocky materials Silicates and oxides of Fe, Mg & Al	Ices and liquids Combinations of C, N_2, H_2, He, O_2, eg H_2O, NH_4	Gases eg H2, H, Ne Ar
Melting 1000 K	300 – 100 K	below 85 K

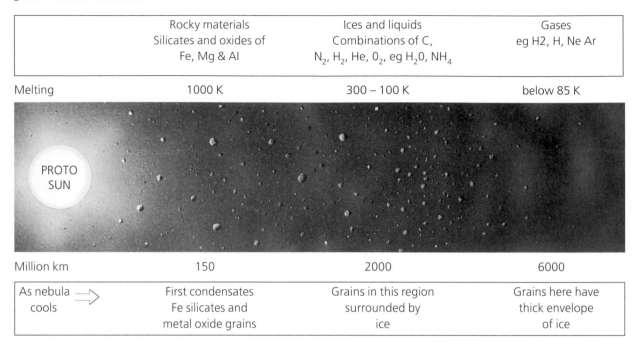

Million km 150	2000	6000
As nebula → cools	First condensates Fe silicates and metal oxide grains	Grains in this region surrounded by ice	Grains here have thick envelope of ice

The effect of temperature upon events in the Solar Nebula. This had profound repercussions for the distribution of rocky, icy and gaseous materials within the Solar System.

helium, neon and argon. These probably never condensed at all. Of the three groups of matter, the first is characterized by very high melting points, around 1250 °C; materials of the second group have lower melting temperatures of around 370–570 °C; those of the third group remain gaseous except under conditions of extremely low temperature. Right from the start, however, the volatile elements were far more abundant than the rest, accounting for perhaps as much as 99.5 per cent of the mass of the nebula.

As matter plummeted towards the central regions of the nebula, the temperatures rose to several thousand degrees centigrade. In these more central regions all matter was vaporized. In the outer regions the temperatures were much lower, probably never exceeding 370 °C, so that dust grains became coated with dry ice (frozen carbon dioxide), water ice and frozen methane and ammonia.

Gradually, as the proto-Sun evolved, the temperature inside the nebula declined, so that gases cooled and solid material began to condense. Because of their high melting points, the rocky materials naturally condensed first, but near the centre of the nebula it was only them that could solidify. In these inner regions the solid grains were large; this was the stuff of the Earth and its near neighbours.

When the Earth's mass had reached about one tenth its present value, gravitational attraction became strong enough to make the incoming planetesimals strike the proto-Earth at very high velocities, high enough to vaporize them upon impact. When this happened, as it did on millions and millions of separate occasions, volatiles were released, and these went to form a primitive atmosphere. Non-volatile elements produced by the same process were condensed and gradually built up an outer layer of molten silicate rock. As a result, the early Earth probably was made up of a cool volatile-rich interior surrounded by a thick ocean of molten rock, perhaps 1000 km deep.

With the passage of time, the decay of long-lived radioactive isotopes locked inside the inner planets gradually heated up their interiors, melting at least parts of them. Thus, under the influence of gravity, the denser materials sank towards the centre of mass, causing chemical redistribution, known as differentiation. Each planet developed an iron-rich core surrounded by layers of mainly silicate material. This was the position reached by the Earth in its primitive stage. Much has happened since that time; the course of evolution of each of the planets taking a somewhat different path.

4

Methods of the earth scientist

Early ideas and methods

While it may be true to say that modern geology did not develop until the late eighteenth and early nineteenth centuries, the pioneers were fortunate in having at their disposal a vast accumulation of observations, speculations and philosophical writings that had amassed during earlier times. During their first dynasty, about 3000 BC, the Egyptians had taken the first faltering steps into the hitherto unknown world of astronomy; however, it was left to the Greeks to take the initiative in explaining some of the perplexing features of the Earth. The fact that it was a Mediterranean race who commenced geological thinking was not mere chance: this region was then, and still is, an extremely active zone of the Earth's crust, giving the Greeks the perfect laboratory in which to study geological processes.

Prior to the time of Pythagoras, about 530 BC, most natural events were attributed to the activities of either gods or mythical heroes. There came a time, however, when superstition and primitive mythological explanations gave way to more carefully reasoned arguments. Admittedly, the seeds of truth which some of the early theories contained were entirely coincidental, but even so we owe them much for the attempts they made to understand the natural processes that were an every-day part of their lives.

The Mediterranean had long been seismically active and there are many accounts of earthquakes in Greek records. Furthermore, the southern Aegean and the south of Italy and Sicily, where the Greeks had colonies, are volcanically active: witness the caldera of Thíra (Santorini) and the line of volcanoes that stretches from Naples to the Aeolian Islands, Sicily and beyond. From the dawn of classical times, people must have lived in awe of fearsome mountains like Etna, Stromboli and Vesuvius, and the trembling ground beneath their feet. Within the region there is also a remarkable diversity of climate and a consequent complexity of geological processes over which

Old Gneiss. *(a)*   Sandstone (Cambrian.)/*b)*.   Lower Quartz Rock. */c)*   Limestone. */d)*   Upper Quartz Rock. etc */e)*

L O W E R   S I L U R I A N.

Title page from Murchison's *Siluria.*

climate has control. There are high mountains such as the Alps and Apennines, glaciers and snowfields, countless fast-flowing rivers, and widespread deposits of rock debris that have been flushed out of the mountains by rivers, eventually to be spread out along the coastline, far from their inland source.

To follow the development of Greek geology would take many pages, but there is not space here to pursue this topic; nevertheless we must note that it was the Greeks who first stumbled upon the real nature of fossils, connected volcanism with earthquakes, and realized that areas now high up in the mountains were once low down beneath the sea. Not all thinking people held such views, however, and very many persisted in believing that fossils were 'sports of nature' – a view that was to survive through mediaeval times. The change toward more modern ideas was a very slow one.

The English antiquary Joshua Childrey appears to have been the first writer to refer to fossils (*Britannia Baconica, 1660*), and Robert Hooke, in his famous work *Micrographia*, published in 1665, produced fine drawings of fossil shells and careful descriptions of them. Buffon, with his *Historie Naturelle*, published in 44 vol-

Geological map of Oxfordshire, England, by William Smith. This map, laid out by John Cary in 1821, is typical of Smith's superb craft. The original scale was 2.5 miles to the inch.

umes (1749–1804), and in particular the volume entitled *Epoques de la Nature* (1779), also deserves mention;

in the latter work we have the very first serious attempt to compute the age of the Earth. Jean Etienne Guettard (1715–1786) was undoubtedly the first of the great mineralogists and he also had a profound knowledge of fossils. His mineralogical map of France and Britain contains a wealth of accurate information, based on painstaking field research.

It was not until 1788 that the next truly major step was taken. This was the publication of James Hutton's paper: 'The Theory of the Earth', in the first volume of the *Transactions of the Royal Society of Edinburgh*. In this ground-breaking article, Hutton expounded the concept of geological cycles during which similar processes operated over very long periods, in much the same way as they do now. The maxim that the present is the key to the past became known as the 'Principle of Uniformitarianism' and marked the beginning of modern geology. In 1795 Hutton published this paper, together with a host of other writings in a two-volume book that became one of the classics of geological literature: it is called *Theory of the Earth with Proofs and Illustrations*.

Mapping rocks

As far as we can tell, the very first piece of geological 'mapping' was accomplished by George Owen who, in words only, described the outcrop of the strata known as the Carboniferous Limestone around the South Wales coalfield. This was published in 1603. The scot, George Sinclair, published a volume entitled *Hydrostatics* in 1672; he appears to have been the first to give an account of a folded and faulted geological structure, giving a clear description of 'dipp' (dip), 'streek' (strike) and 'crop' (outcrop) into the bargain.

The term outcrop refers to any exposure of bare rock. If this is a sedimentary rock such as a marine sandstone, then this will doubtless be stratified since it will be composed of rock fragments derived from the land by erosion, then transported and deposited in water. It will then have been buried, compressed and finally brought to the surface again by erosion. Stratification means that a rock has been laid down as a horizontal (or subhorizontal) layer (stratum) composed of individual grains which subsequently became coherent. The coherence may have been achieved largely by pressure or by the precipitation of a chemical cement between the grains. The study of stratified rocks forms the basis of geological mapping, while analysis of the fossil remains contained in such strata is termed palaeontology, a discipline which enables geologists to place the geological history of an area into the broader context of Earth history, and allows biologists to study evolutionary trends in the biota.

Strata may not, of course, remain horizontal. The Earth is a dynamic planet and periods of internal upheaval may deform rocks, rucking them up into folds, tilting them or even breaking them apart. Folded strata are common and if deformation is very intense, it may cause them to turn upside down. This produces a very complex situation which only detailed study and mapping can elucidate. A geological map endeavours to give a two-dimensional picture of a three-dimensional situation. Thus, where strata are inclined to the horizontal, the annotation of the map will indicate the inclination or 'dip' of the beds from place to place. The trace of a line drawn perpendicular to dip is termed the strike. Dip, strike and outcrop are three fundamental geological terms.

Rock outcrops may well be widely scattered over a region; between them may be soil, buildings, vegetation or water. One of the prime tasks of the field geologist is to correlate beds located in widely separated areas, perhaps on a different continent. To do this he or she needs not only to study the nature and disposition of the rocks but also of the contained fossils. Fossils are the remains of dead organisms and it is now well understood that organic life has evolved in particular ways. Thus, certain kinds of organisms

Looking down the strike of a series tightly folded vertical Silurian strata, Aberystwyth, Wales. Careful mapping of the dip and strike of folded strata of this kind gradually helped geologists to understand how mountains are formed and how the Caledonian mountain-building cycle affected Britain.

lived only during relatively restricted intervals of time; specific genera or species evolving in particular and unique ways. Palaeontologists now have a pretty good idea about what kinds of organisms lived in the many kilometres thickness of strata which comprise the upper part of the Earth's crust, and of the time spans and types of geological environment in which they thrived. Not surprisingly, each set of rock strata has its own particular fauna and/or flora.

Today there are geological maps of virtually all of the Earth's surface, even the floors of the deepest oceans and the tops of the highest mountains. Such maps also exist for other planets and their moons. Modern technological developments could not have been predicted by the early pioneers: who could have predicted that orbiting satellites would now be making such a major contribution to our knowledge of Earth and the planets? In the next decade it may well be that orbiting space-stations will be making the

next strides forward in the type and quality of terrestrial mapping.

Going back a little in time, perhaps the finest of the early geological maps were produced during the latter part of the eighteenth century, in particular by the famous engineer-cum-surveyor-cum-geologist, William Smith. Smith was a mineral surveyor who, in his travels, visited many mines and quarries. He began to recognize the regularity of the strata he encountered, and the great value of the fossils they contained. When subsequently he became involved in surveying for the great canal network that was built in southern Britain in the late eighteenth century, he came quickly to recognize the different strata in widely scattered regions, not only because of their colour or mineral content, but also by the fossils preserved in them.

Between 1794 and 1809 Smith collated a series of geological sections across various parts of England.

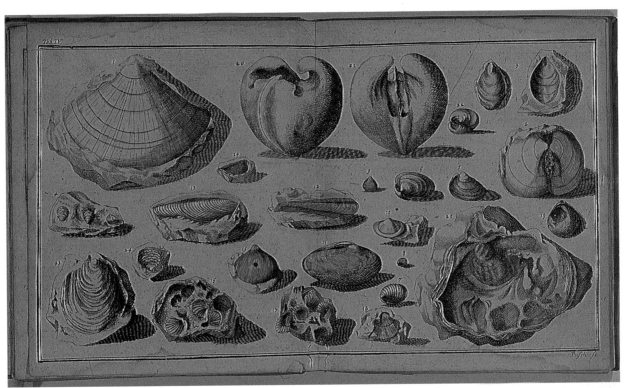

Early drawings of marine fossils. This particular set appeared in Johann Jacob Baier's *Oryctographiae Noricae*, published in 1712.

His immaculately drawn section of the strata between Bristol and Norwich appears to have been the first of its kind. Even more impressive were his fine geological maps, particularly the one published in 1799 and titled: *Geological Map of the country around Bath*. By 1815 he had published a series of fifteen maps that covered most of England and Wales. These have become classics and the originals fetch very high prices in antiquarian bookshops. His 1801 map was the first accurate representation of the geological strata of England and Wales on a single sheet. We owe much to this great man, not only because he achieved so much, but also because he set such a high standard for later workers to emulate.

The fossil record – 1

Most ancient writers explained the preservation of fossils in rocks as the result of great natural catastrophes which effected a change in the relative positions of land and sea. With the dawn of the fifteenth century and the revival of learning that accompanied this, came three centuries of vigorous debate: were fossils simply 'sports of nature', did they originate in some kind of magical fluid (*vis plastica*), or were they remains of past life forms?

The fifteenth century genius, Leonardo da Vinci, was among the first to realize that fossils showed that the strata containing them had once lain beneath the sea. A similar conclusion was reached by the Frenchman, Bernard Palissy, (c.1510–1589) who published drawings and descriptions of petrified wood, together with fossilized fishes and molluscs. For his pains he was verbally attacked by his contemporaries and soundly denounced by the Church.

The English zoologist Martin Lister (1638–1711) had an excellent knowledge of fossil shells, while the Celtic scholar and antiquary, Edward Lluyd (1660–1709) was another early writer who had a detailed knowledge of fossils. In one of his beautifully illustrated books he has plates of over 1000 fossils kept at the Ashmolean Museum in Oxford. In a letter to the famous biologist, John Ray, he suggested that such fossils might be 'partly owing to fish-spawn received into the chinks of the earth in the water of the Deluge' (see Ray, *Three Physio-Theological Discourses* (1702) p. 190).

During the seventeenth century the Dane, Nicolaus Steno, and the British scientist/philosophers Robert

Hooke, John Ray, John Woodward and others guided scientific thought towards more factual lines, so that by the middle of the ninenteenth century, no true man or woman of science still adhered to the old notions. Robert Hooke (1635–1703) was one of the most brilliant men of his time and apparently was the first to propose that fossils might be used in revealing the Earth's past history. Inclined to the view that fossils of unknown forms were extinct species that had been annihilated by earthquakes; he also observed that some fossils appeared to be restricted to specific localities, while the presence of others indicated that while they lived, the climate must have been different from that of contemporary times.

No sooner had some of these new ideas been developed than new social developments spawned the Diluvialinists. Blessed by the Church, adherents' activities slowed down progress for many years. Eventually, however, with the emergence in the late eighteenth century and early nineteenth centuries of such men as William Smith, Georges Cuvier and Alexandre Brongniart, modern palaeontology gathered momentum. In 1799, Smith drew up a table of British strata from the Coal Measures to the Chalk (the former a lithological division of the Carboniferous System in Great Britain, the latter, of the Cretaceus), and a little later published his famous *Strata Identified by Organized Fossils* (1816), and also his *Stratigraphical System of Organized Fossils* (1817). These works, once and for all, put stratigraphical palaeontology on a firm footing.

The realization that fossils were of vital importance in determining the rock succession was to have a profound effect on geology. Working side by side, Cuvier and Brongniart studied the rocks of the Paris Basin and gradually established that there was a systematic difference between the fossils they collected at different horizons. They wrote: 'The means which we have employed, among so many limestones, for the recognition of a bed already observed in some distant quarter, has been taken from the nature of the fossils contained in each bed. These fossils are generally the same in corresponding beds, and present tolerably marked differences of species from one group of beds to another. It is a method of recognition which up to now has never deceived us' (see *Essai sur la Géographie Minéralogic des Environs de Paris, avec une Carte géognostique et des Coupes de terrain*, 1911)

Cuvier, probably remembered most widely for his applications of comparative anatomy to fossil vertebrates, and his friend Brongniart, had at last established the principles of stratigraphic palaeontology. It was a major landmark in the development of geology. Their French contemporary Jean-Baptiste de Lamarck, who was studying the invertebrate fossils in his region, was the first of the true evolutionists. He had a wide knowledge of fossils and was far more advanced in his ideas than Cuvier, for instance. He firmly adhered to the idea that species were descended from their ancestors, but showed modification as time proceeded. Evolutionary palaeontology can be said to have started not with Cuvier and Brongniart, but with Lamarck.

As more and more scientists contributed to palaeontology, their worked confirmed Cuvier's view that fossils showed a progressive change with time. Furthermore, as collecting spread more widely through the European continent, similar successions of fossil organisms were uncovered in widely divergent places. These pioneering palaeontologists gradually realized that many fossil species, genera, or even families, were limited to fairly narrow stratigraphical intervals. Additionally, they discovered that some stratigraphically limited groups were widespread geographically. This led to the statement of the Principle of Faunal Succession: like assemblages of fossils indicate geological ages of the same order for the rocks that contain them. It also led to a standardized scheme of stratigraphical terms.

Since the time of Cuvier and his contemporaries,

(a) An ammonite infilled with crystalline silica. (b) Internal cast of a gastropod. (c) The headshield of a trilobite.

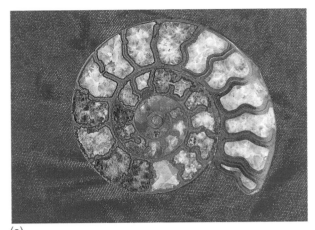

(a)

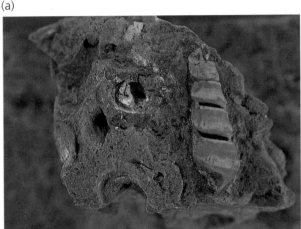

(b)

(c)

generations of palaeontologists have assiduously collected and collated fossils from nearly every corner of the globe. The oldest strata that contain abundant fossils have an age of around 570 million years, and belong to what geologists know as the Cambrian System. This point in time also marks the commencement of what is termed the Phanerozoic Era – the period during which life flourished on Earth. It should be noted, however, that older rocks than Cambrian strata also contain the remains of more primitive life forms, often poorly preserved, usually sparsely distributed, but nevertheless recording the presence of life on Earth as long ago as 2500 million years.

Early palaeontology was largely descriptive; today it has undergone a revolution. The twentieth century palaeontologist is largely concerned with refining ways in which fossils can be used to date strata relatively, in understanding the processes which result in organic evolution, and in the economic applications of fossil studies. The tools of the trade are not only the hammer and chisel, but computers and electron microscopes.

The fossil record – 2

The process of fossilization is very complex and may involve several stages, beginning with the death of an organism and ending with its discovery as a fossil. Organisms with mineralized shells have a far greater chance of survival than those with flimsy or delicate skeletons, or none at all; thus there is a considerable imbalance in our knowledge of ancient life through fossils.

What the palaeontologist refers to as 'body fossils' are the relatively tough skeletal remains which survive intact after the soft tissues have decayed. Such are the shells of molluscs and crustaceans, and the bones of vertebrates. Even these may come under attack: marine shells may be bored into by other

organisms, or broken up while being trasnported across the sea floor by currents; they may even be dissolved during chemical reactions that occur after burial.

Those parts which do survive must eventually become buried in sediment for them to survive further degradation, thus animals which live within the sediment itself are at a distinct advantage. They have a substantially greater preservation rate than free-swimming forms or bottom-crawlers. The survival of plant tissue is an equally chancy business: very delicate vegetable remains may be preserved if they are buried speedily, for instance by volcanic eruptions, or by sand and mud brought down during floods or earthquakes.

After burial, organic matter will be subject to compaction due to the weight of the overburden; this, naturally, increases with depth. Some types of skeletons will be crushed flat during this process, others may survive tolerably well. Some replacement or partial chemical alteration of the original material may take place; chemical reactions being catalysed by the increase in temperature and pressure experienced by buried strata. Such alteration is very common among relatively soluble calcareous organisms.

Calcareous organisms are composed of calcium carbonate ($CaCO_3$) which occurs principally in two forms, calcite and aragonite. These differ in their internal structure. The latter form is particularly unstable and usually is dissolved away or is converted into the more stable form, calcite. Aragonite thus becomes less often encountered as we move down through the rock succession into older and older horizons. Sometimes calcite or aragonite may be totally removed in solution, which leaves a cavity. In time this may be filled by more stable minerals, giving rise to moulds or casts of the originals. Such modes of preservation frequently reveal delicate anatomical features that otherwise would not be visible. At other times the calcium carbonate may simply be replaced, crystal for crystal, by silica or iron pyrites.

A relatively small proportion of animals actually have skeletons composed of silica (SiO_2), examples are the sponges. Others may be composed in part of calcium phosphate – like the conodonts – which are preserved in the mineral apatite, a species very resistant to either solution or erosion. Whatever the composition of the skeletal material, it has a substantially

greater chance of survival than the soft tissue. Not surprisingly, therefore, when a major discovery of some delicate soft tissued organism is made, it often makes press headlines. One example was the discovery of a frozen baby mammoth in Siberia; this had lived over 10 000 years ago.

Trace fossils represent preserved tracks or other markings made in the soft sediment of the sea floor by animals which lived either on the sea-floor or just beneath its surface, in the underlying mud or silt. Frequently it is impossible to determine which particular kind of organism made such impressions, but in certain instances more definite identification can be made. This is particularly true of the trilobites, whose footprints have been documented from the Lower Palaeozoic rocks of Wales and elsewhere. The more common trace fossils represent tracks left by activities such as burrowing, crawling, feeding or resting. Their greatest value is as indicators of water oxygenation levels and the diversity of the original biota; they have some value also as indicators of the environment in which the organisms lived.

Fossil plants found in strata associated with coal deposits are usually severely compacted and the internal features destroyed; however, they are often beautifully preserved as impressions. Occasionally the internal anatomy is revealed because the plant was invaded by mineral solutions before it could be compressed and crushed. The silicified tree trunks of famous Petrified Forest of Arizona are examples of this kind of preservation.

Although our knowledge of the fossil record is substantially more complete than in the days of Smith and Lamarck, there are still some large gaps in the record. Some of these are due to the almost complete destruction of some delicate life-forms; others remain because palaeontologists have yet to study and collect from all available localities.

Two kinds of seismographs: (a) pendulum-mounted instrument that records horizontal movements; (b) spring-mounted instrument that records vertical motion.

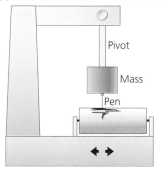

(a) Pendulum-mounted Seismometer

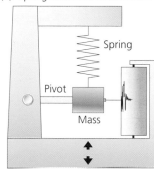

(b) Spring-mounted Seismometer

Methods of the geophysicist – 1

In their quest to perceive what lies at greater and greater distances from the Earth, astronomers have invented successively more powerful telescopes which have exploited different parts of the electro-magnetic spectrum. The geophysicist has pursued a similar quest, but in the opposite direction: in his case the need is to understand what lies at greater and greater depths inside the Earth. To achieve this it is necessary to use not visual wavelengths, but longer ones, i.e. sound or *seismic* waves. Fortunately the Earth's rocks allow passage of seismic waves and the recording of such vibrations has formed the basis not only of our understanding of the planet's internal structure, but also of several widely-used methods of prospecting for oil and gas.

Earthquakes produce seismic waves which travel through solid rocks and whose nature and arrival time at the surface can be recorded very accurately on delicate instruments called seismographs. These were invented in the late nineteenth century and provided scientists with confirmation that both compressional and shear waves are generated by earthquakes. Further studies revealed also that each type of wave was not transmitted through all parts of the Earth's interior with equal facility. Interpretation of these findings led to our first reliable ideas about what lies inside the Earth (see also Chapter 6).

A seismograph is an instrument for recording any change in distance between itself and the rocks beneath it. Theoretically the device is not itself affected by movements inside the Earth, but in reality this cannot be achieved. Instead, an approximation to this condition is realized by mounting a small mass on either a pendulum or the end of a delicate spring, so that when the ground moves, the mass remains still because of its inertia, while the Earth is displaced relative to it. The displacement is recorded on the seismometer by a pen which records the motion on a rotating drum – the seismograph is the result.

Modern exploration geologists generate their own mini-earthquakes to gain information about subsurface strata, thereby increasing the efficiency with which they can predict the presence of viable reservoirs of oil or gas. A series of small seismic waves is generated, sent down into the crust, reflected by the subsurface strata and returned to the surface to be recorded on a detector. Generally a series of detectors is used, these being arranged in a straight line which stretches away from the seismic source. On land the energy source is sometimes an explosive, but for maritime surveys it is normally a kind of air gun.

To record the reflected waves, the large seismometers needed for monitoring earthquakes are unnecessary; instead small recorders called geophones (on land) or hydrophones (on the sea) are used. During maritime surveys these usually are towed behind a research vessel as a line called a 'seismic streamer', in a depth of between 5 and 15 m of water. Computers process the received signals, apply certain corrections and issue a seismic reflection profile. Such a profile reveals not only the subsurface structure but also gives information about the physical properties of the hidden rocks.

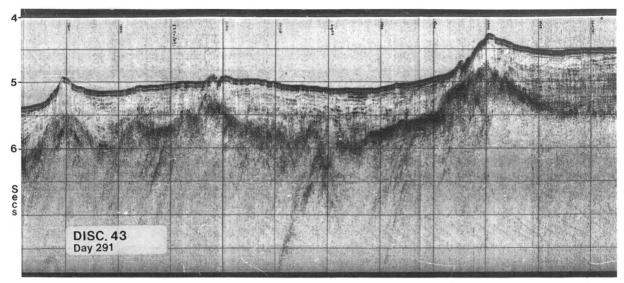

Seismic reflection profile across part of the North Atlantic Ocean. The subsurface structure and sedimentary layering are clearly revealed in this traverse. [Photo: Mark Noel]

Another geophysical technique uses a logging tool called a *sonde*, which is lowered on a steel cable into a borehole. Called the electrical wireline logging technique, it is widely utilized in the hydrocarbon industry, where decisions to exploit or not exploit a new field are based on interpretation of the well 'logs' so obtained. The sonde itself may contain several different instruments which make a variety of measurements. Their measurements normally will lead to a knowledge of rock porosity, the ease with which fluids can travel through rock pores (permeability) and the nature of fluids trapped within the pores themselves and may also include readings of natural gamma ray activity. The results are plotted as a curve of measurement against depth.

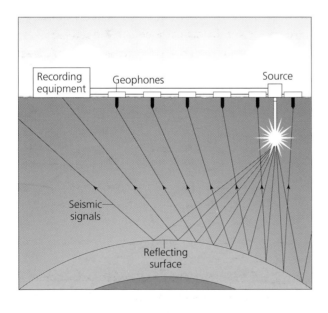

Seismic technique for probing subsurface geology. Small explosions generate sound waves which are bounced off rock layers beneath the surface and bounced back to geophones.

Methods of the geophysicist – 2

About a century and a half ago, British surveyors involved in the mapping of the Indian subcontinent encountered an inconsistency in their survey which led to an appreciation of one of geology's fundamental concepts, *isostacy*. The details of the discovery are sufficiently interesting to describe in a little detail. During the survey the distance between Kaliana – which lies about 150 km south of the Himalayas – and Kalianpur – some 600 km further south – was determined in two very precise ways: by measurements over the land surface, and by reference to astronomical observation. Strangely, the results disagreed by some 150 m, which may appear trivial, but to the surveyors was an unacceptably large discrepancy. In trying to explain away the difference, it was proposed that the plumb-line used in the survey was tilted towards the mountains because of the gravitational attraction they exerted on the plumb-bob.

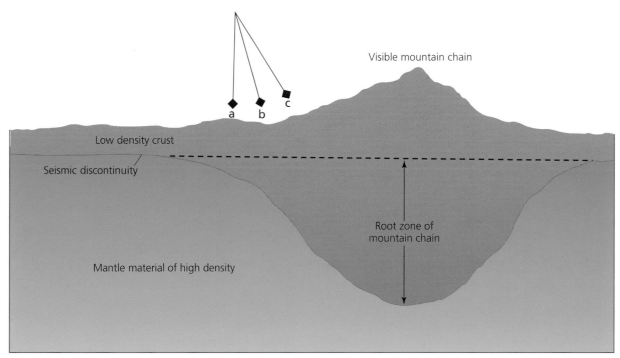

Isostacy: mountain masses deflect a pendulum away from the vertical, but not as much as might be expected. In the diagram, the vertical position is shown by (a); if the mountain were simply a load resting on a uniform crust, it ought to be deflected to (c). However because it has a deep 'root' of relatively non-dense rocks, the observed deflection is only to (b)

Calculations were made but the theoretical effect indicated the discrepancy should have been greater than it appeared to be – 450 m! The problem was compounded rather than solved, and remained so for some time. Then, in 1865, Sir George Airey, the Astronomer Royal, came forward with the suggestion that the enormously massive Himalayas were not actually a part of the rigid crust but were buoyed up on a layer of denser rock, rather as the hull of a ship in water. This meant that the excess mass of the mountain chain above sea level was compensated for by a mass deficiency in the underlying 'root' zone. His explanation became accepted by geologists and forms the basis of the principle of isostacy. Application of this principle has played an important part in our understanding of the behaviour of planetary crusts and lithospheres.

The same principle forms the basis of another common method of geophysical measurement, the gravity survey. The objectives of this are to measure the mass variations that occur inside the Earth, and are achieved with the use of an instrument called a gravimeter. A small device no larger than a wine bot-

tle, it consists of a weight attached to a spring that stretches or contracts in response to an increase or decrease in gravity from place to place. Modern instruments can detect variations as small as one hundred millionth of the Earth's gravity. The standard unit of gravity measurement is the milligal, which is equal to a gravitational acceleration of 0.001 cm/s².

As with many of the modern geophysical techniques, costs involved in their development have been borne largely by the oil industry which, about 40 years ago, realized that the buried geological structures which had the capacity to trap oil often produced variations in the normal gravity field. These could be detected by sufficiently sensitive recorders. The *gravity anomalies* are caused by a change in the subsurface mass, due to the presence of such structures as salt domes

The principle behind the surveys is that any anomalous mass will have an effect upon the Earth's gravitational field and that careful surveying of the gravity may be used to reveal details of the subsurface structure. Before a profile can be useful, however, a number of corrections need to be made, which arise largely because the Earth is not a perfect sphere. The polar flattening, produced by the centrifugal forces that operate in the rapidly-rotating planet, means that the force of gravity is less at the Equator than at the poles. A correction therefore has to be made for latitude. Furthermore, a gravity station set up on a hill is further from the Earth's centre than one at sea level, and will thus give lesser measurements. This effect must also be taken into account. Finally, if the profile eventually derived is to be properly representative of the subsurface geology, care must be taken in assessing the gravity contribution of all near-surface masses of rock, correcting accordingly. Having made all of these corrections, gravity surveys become a very potent tool in the elucidation of the Earth's subsurface geology. This is true also of other planets, and gravity data have been collected by spacecraft on most of recent planetary mapping missions, the last 243-day mapping cycle of the Magellan probe, which orbited Venus, was specifically a high-resolution gravity programme.

Techniques of the geochemist

The Earth's crust – the part of the planet which geologists can study at first hand – is composed of aggregations of different minerals, mainly silicates, which constitute rocks. These may contain varying proportions of a small number of major elements: silicon, oxygen, aluminium, calcium, iron, magnesium, titanium, manganese, sodium, potassium and phosphorous. These combine into complex atomic structures to form a small number of silicate mineral groups, e.g. olivine, feldspar etc. A very small proportion of the atomic sites may be taken up by rarer elements, called trace elements which have to fight to enter the mineral structures but will do so if the precise physical and chemical conditions are right. Typical trace elements are barium, rubidium, strontium, chro-

mium and vanadium. Often there may be chemical substitution of some major elements by trace elements, for instance nickel may replace some of the magnesium and iron in the mineral olivine, while barium may replace some of the potassium in feldspars. The behaviour of all these elements is of extreme importance to the petrologist and geochemist, since it allows them to establish such factors as the state of oxygenation , the temperature and the pressure within the system being studied, which could be magma, sea-water or a pore fluid, to name just three examples.

Early mineralogists identified minerals by studying their physical properties and analysed their chemical make-up by 'wet' gravimetric techniques. Later the polarizing microscope provided an additional and very powerful diagnostic tool, extending the study of minerals' physical and optical properties. In the early twentieth century X-rays were used to investigate crystal structure, and subsequently there was the introduction of more sophisticated techniques, for instance those based on atomic absorption, X-ray fluorescence and the electron microprobe. One of the limitations of the polarizing microscope is that, in accordance with the rule that resolution is limited by the wavelength of the radiation being used, there is a lower limit to the size of objects resolvable in such instruments, this being somewhere in the region of one half of a micron (10^{-6} m).

Such a constraint is not experienced with the more modern instrument, the electron microscope, where far shorter wavelengths are utilized. Electron beams can be reflected, refracted and diffracted, just like light can; however, their wavelength is of the order of 10^{-12} m. Because they are charged particles it is possible for them to be focussed into a beam and magnified by varying the strength of the current in an electromagnetic lens. The electrons themselves are produced by a heated filament, the whole system being operated under vacuum conditions. The sophistication of such a microscope means, of course, that it is very expensive, and can only be afforded by a small number of scientific institutions.

Another modern technique – optical emission spectrography – depends upon the spectral analysis of radiation generated by the excitation of atoms when they are strongly heated. The radiation produced when a rock sample is heated to a high temperature is passed through a prism, is refracted and dispersed according to wavelength. A typical optical emission spectrum of an element consists of a series of lines which correspond to radiation of different specific wavelengths. The intensity of the lines in the spectrum can also be measured and this gives an indication of the concentration of individual elements in the sample.

X-ray fluorescence spectrometry (XRF for short) is another technique for determining the chemical consititution of rocks and their contained minerals. It is very widely used and is a relatively cheap and quick method of analysis. It depends upon the principles of both X-ray diffraction and emission spectrography:

the atoms in the sample are excited by a primary X-ray beam which causes them to emit energy at X-ray wavelengths (very short wavelength). The radiation so produced is characteristic for each chemical element and the intensity is proportional to the concentration of the element in the sample.

One of the most sophisticated developments has been the invention of the electron microprobe, pioneered by J. V. Long at Cambridge in the 1950s. This sophisticated instrument can, from a small polished slice of a rock sample, analyse the chemical composition of individual mineral grains; indeed it can measure the variations in chemistry across a single grain or crystal. This is achieved by focussing a high-energy electron beam on a polished, carbon-coated sliver of rock. The X-rays from the sample can then be measured in much the same way as in XRF analysis. One of the joys of this instrument is that it is possible to view the sample in the eyepiece of a polarizing microscope, thereby allowing the operator to select the optimum position within a grain to be analysed. The probe is an extremely expensive commodity!

Part two

The Earth's heat engine

Solfataras emitting steam and gas on the rim of the Fossa Crater, Vulcano, Aeolian Islands, part of an active island arc.

5

The Earth's heat engine

The Earth heats up

By making very reasonable guesses about the early temperatures and pressures inside the Earth and about the thermal condition of radioactivity, it is possible to compute what may have occurred during the first one billion years of the Earth's history. Most such calculations indicate that accretional and radioactive heat was unable to escape fast enough to prevent the Earth heating up, thus, by about 3500 million years ago, the internal temperature was sufficiently high to cause melting of metallic iron. This had a far-reaching effect upon the planet's subsequent evolution.

Many factors suggest that the Earth began to accrete after metallic and silicate grains had condensed from the cooling Solar Nebula. It seems likely that metallic particles containing iron were preferentially accreted due to the greater ductility and higher density of iron-rich molecules compared to the more brittle silicates; the latter accumulated later to form the Earth's mantle. Evidence provided by meteorites indicates that in addition to small grains there were larger ones that existed very early in Solar System history. Such planetesimals probably were at least 100 km in diameter, and also contributed to the construction of the primaeval Earth. Accepting this to be so, it would not be long before a dense core began to form. Core formation accompanied the growth of all the planets, but only the inner ones contained dense elements such as iron and nickel (the outer planets saw metallic hydrogen sinking towards the centre).

As the Earth grew, so gravity increased. Calculations confirm that sinking of iron could have commenced after only one eighth of the Earth's final mass had accumulated, in which case core formation must have started before the whole Earth had accreted. Once the separation of iron and silicates began, Earth would have heated up much more rapidly because gravitational potential energy would have been released. Very rapid accretion would lead to very rapid heating up, since the rate of radiation of thermal energy into space would have been quite slow. There is every reason to believe that during this phase the Earth's internal temperature rose by about 2000 °C, sufficient to trigger wholesale melting and the setting into motion of an even more efficient process for the redistribution of the planet's chemical ingredients: convection.

Stirrings inside a planet

The formation of the core marked a new phase in Earth's development; it became a layered body inside which the denser components sank towards the centre while the lighter ones accumulated nearer to the surface. In other words, Earth underwent a primary *chemical differentiation*. Estimates for the time of core formation range between 4500 and 3700 million years ago. At this stage the internal heat could escape only by conduction – a very slow process – consequently the temperature rose quickly until the Earth may have become largely molten. Now, in any body of liquid which is hotter at the bottom than at the top, the hotter material expands and floats upwards over the denser, cooler material above it until it, in turn, cools, becomes more dense and then sinks again; i.e. it convects. By this process heat was much more efficiently transferred towards Earth's surface and dissipated into space.

As time passed the Earth cooled sufficiently for its outer regions to solidify, although the core remained molten. Indeed, the outer part of the core remains so today and it appears that even 4500 million years is too short a period to allow the core to have 'frozen'. The relatively rapid rotation of the Earth doubtless has helped in keeping the outer core molten, but it is also a function of the planet's not inconsiderable mass.

An analysis of seismic waves and how they behave as they pass through the body of the Earth indicates that the Earth evolved such that it has a solid inner core mantled by a liquid outer core; the two layers extend from about 2900 km below the surface to the planet's centre. Encircling the core is a less dense layer of solid silicates – the mantle – which is trapped between the outer core and the thin, silicate crust, the latter being composed of relatively non-dense silicates with quite low melting points. Internal convection persists to this day, as is shown by the slow movements of the outer skin of the Earth – a response to the motion that is taking place within. Yet, how can this be so if the mantle layer is solid?

Suprisingly enough, under certain conditions, even solid rock can be set in motion. The realization that warm but solid substances can convect came to the German-born US seismologist, Beno Gutenberg in 1914. His notion of solid-state convection revolutionized geophysics. It is now accepted that if rocks become very hot they expand and become less dense; moving upwards towards a planet's surface, very slowly, in a motion akin to convection; this is called 'high-temperature-creep'.

Abundances of the elements

The chemical elements residing in the Earth formed compounds, each with different melting points, densities and chemical affinities. It was these differing properties that governed their subsequent redistribution. For example, the feldspar group of minerals – silicates of aluminium, potassium, sodium and calcium – have relatively low melting points and are of low density. These would be expected to rise towards the Earth's surface during differentiation – as they did – and become the most abundant common mineral group in the crust. The mantle, on the other hand, became a reservoir for the less easily melted, denser silicates of magnesium and iron, such as the pyroxenes and olivines.

Of the less abundant elements, some heavy ones, like platinum (Pt) and gold (Au) sank towards the

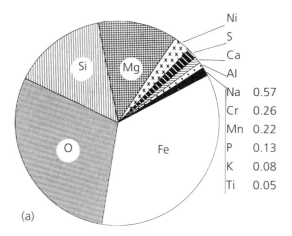

Na	0.57
Cr	0.26
Mn	0.22
P	0.13
K	0.08
Ti	0.05

(a)

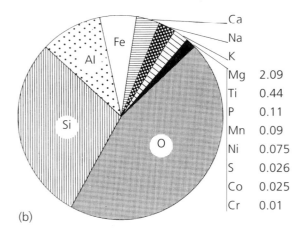

Mg	2.09
Ti	0.44
P	0.11
Mn	0.09
Ni	0.075
S	0.026
Co	0.025
Cr	0.01

(b)

Pie diagrams showing the elemental abundances in the Earth as a whole (a) and the Earth's crust alone (b).

core; however, this was not simply because of their density but also because they show little affinity for either silicon or oxygen, and therefore do not enter the lattices of silicate minerals so typical of the crust and upper mantle. On the other hand, some heavy elements, such as uranium (U) and thorium (Th) which have affinities for both oxygen and silicon, accumulated in the crustal layer, despite their high density. Surprisingly perhaps, gravity played but a secondary rôle in determining how the chemical elements were to be redistributed.

One interesting (and important) result of the chemical fractionation was that radioactive elements like thorium (Th), rubidium (Rb) and neodymium (Nd) were gradually concentrated in the Earth's outer layers, in complete contrast to their originally even distribution within the Earth. In the early part of Solar System history, the decay of radioactive elements significantly raised the temperature of the whole Earth; however as this radioactive fuel started to accumulate in the crustal regions, the heat produced was lost more readily to space. This provides an example of how chemical differentiation had the effect of retarding the action of the Earth's 'heat engine' – allowing energy to leak more quickly away.

The elemental redistribution effectively separated those elements with an affinity for oxygen and silicon (i.e. silicate), termed *lithophile* elements, from those with an affinity for metallic iron – *siderophile* elements – and sulphur (*chalcophile*). Subsequent degassing (see the next section) separated off the *atmophile* elements.

The movement of volatiles

The primaeval planetesimals from which the Earth and other planets accreted could not have held on to their volatile constituents for ever. Indeed, because of their low masses they would have had only a weak gravitational pull, meaning that the faster-moving elements (hydrogen, helium, etc) could escape easily into space. The Earth's primitive atmosphere must therefore have been produced by degassing of the interior. Since it is now generally accepted that the Earth was formed by accretion of more-or-less cold fragments, there is little alternative to deducing that it was the Earth's heat engine that caused these more volatile elements to move towards the surface. Since they ended up largely in the atmosphere or hydrosphere, they are called *atmophile* elements

Silicate minerals which are common in the Earth's crust, e.g. micas and amphiboles, contain both hydrogen and oxygen, in addition to elements such as silicon, iron and magnesium; the hydrogen and oxygen are bound into crystal lattices as the hydroxyl molecule (OH). As the Earth heated up and chemical differentiation progressed, molten rock (*magma*) was generated by the partial melting of existing solid materials, and, together with water and other volatiles, rose towards the surface. If it reached the surface, volcanic eruptions ensued, and the volatiles escaped into the atmosphere, forming clouds of hot vapour. Assuming that the past frequency of eruptions was similar to today's, the predicted numbers of magmas reaching the surface in the past would easily have contributed sufficient water to fill the ocean

Steam escaping from vents in the jungle at Kawah Kamocang, Western Java. This is a highly active part of the Indonesian island arc.

Fumaroles emitting steam, sulphur dioxide and hydrogen sulphide at Kawah Mas, Gunung Papandayan, Western Java.

basins during geological time. In fact, it is very likely that such eruptions would have been far more frequent in the geological past than now, making the task of generating Earth's oceans even easier.

The primaeval atmosphere must have been quite unlike today's. Modern volcanoes exude gases in large volumes and studies of these suggest that in the early days water vapour, together with carbon dioxide (CO_2), carbon monoxide (CO), nitrogen (N_2), hydrogen chloride (HCl) and hydrogen (H_2) would have been most abundant. Of these the very light gas, hydrogen, must have escaped very quickly into space, while some of the water vapour in the upper layers of the atmosphere would have been broken down into hydrogen and oxygen by the action of sunlight (photodissociation), the former escaping, the latter mostly combining with gases such as methane (CH_4) and carbon monoxide to form water and carbon dioxide. The production of large quantities of free oxygen, one of the principle components of today's atmosphere, was a long way off, and had to await the development of plant life.

Heat and volcanism

The heat engine, fuelled early in Earth's history by accretional energy, then by the decay of short-lived radioactive nuclides, and later by long-lived radioactive materials, constantly causes modification of the surface. The cycle of heat production, its transfer to the surface, followed by its eventual dissipation into the atmosphere and/or space, produces fundamental changes in the surface; this has been true of all of the terrestrial planets, and is still true for the Earth, Venus and Mars. Earth, out of all the terrestrial worlds, has remained the most dynamic and shows the most diverse pattern of surface change: volcanism, seismicity, tectonism, weathering and erosion all are taking place, constantly.

Because of their relatively small masses, both Mercury and the Moon suffered rapid heat loss early in their histories, with the result that their surfaces were soon protected from internally-driven modification by thick mantles and crusts which became cold and rigid relatively quickly. The intercrater plains of Mercury, however, appear to have had substantial

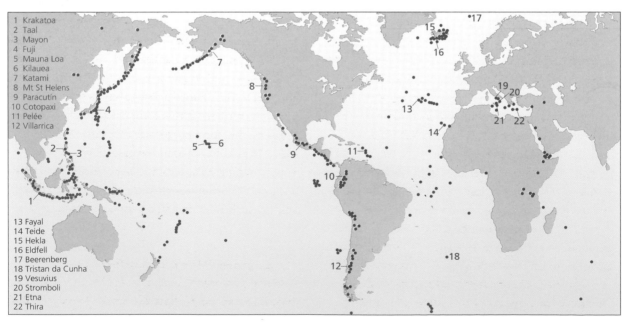

1 Krakatoa
2 Taal
3 Mayon
4 Fuji
5 Mauna Loa
6 Kilauea
7 Katami
8 Mt St Helens
9 Paracutín
10 Cotopaxi
11 Pelée
12 Villarrica
13 Fayal
14 Teide
15 Hekla
16 Eldfell
17 Beerenberg
18 Tristan da Cunha
19 Vesuvius
20 Stromboli
21 Etna
22 Thira

Distribution of the world's active volcanoes. The strong concentration along continental margins and mid-ocean ridges is clearly shown.

Night-time eruption of Stromboli volcano, Aeolian Islands, Italy. Photograph taken in late May 1996.

input from an early phase of volcanism which likely ended over 3500 million years ago. The Moon's main phase of volcanism effected the emplacement of the basalt-like lavas which filled ancient impact basins to produce the lunar maria. This occurred between about 3900 and 3200 million years ago. Both earlier and later volcanism also occurred, but was of a more restricted kind. Mars still shows modifications due to interactions between the atmosphere and surface, but cooled more rapidly than the Earth, with the result that volcanic activity and tectonics appear to have died out many hundreds of millions of years ago, maybe longer. Venus, on the other hand, a planet of much the same mass as the Earth, shows ample evidence of both volcanic and tectonic modification of its surface into quite recent times.

That there is sufficient heat inside the Earth to melt pockets of mantle material, is proven by the frequent outbreaking of volcanicity. This may manifest itself in the pouring out of molten lava, by the production of massive eruption plumes of molten magma and rock debris which rise high into the atmosphere, by huge gas explosions, or by the injection of large volumes of magma into the upper crust as small intrusions such as dykes and sills, which are only discovered after many years of erosion or by tectonic collapse events. Whichever is the case, there is an underlying factor which is of great importance: as magma rises through the crust, the dissolved gases it carries with it expand as the load pressure decreases. Eventually a point may be reached when the pressure exerted by the dissolved gases exceeds that of the rock overburden; at this point the gases come out of solution and form bubbles; the magma is said to *vesiculate*. The expansion of these gases is vital to the uprise of magma and its manner of escape has a far-reaching effect on the style of eruption that ensues and of the kind of volcanic structure which results.

Clues from volcanic landforms

Each of the terrestrial planets, the Moon and Jupiter's innermost moon, Io, have experienced silicate volcanism at one time or another. Additionally, research into geological activity on the other outer planet moons has revealed that eruption of cool fluids has taken place there also, being the result of cryovolcanic activity, or *cryovolcanism*. Earth and Io currently are volcanically active, and analysis of Magellan imagery of Venus suggests that it too has been volcanically resurfaced quite recently and may become more active again in the future. The volcanic landforms encountered are very diverse, and so it is appropriate that we enquire what factors control their location and morphology. An understanding of these will inevitably provide many clues regarding the internal workings of the Earth and its planetary neighbours.

The form of a volcanic landform depends predominantly on the chemical composition of the magma produced during a volcanic event, how much of it

there is, and exactly how it is released at the surface. Lavas containing small amounts of silica, like basalts (which typically contain 45–50 per cent SiO_2), are generally rather fluid, having low viscosity; silica-rich lavas, such as dacites and rhyolites (65–75 per cent SiO_2), tend to be very viscous and flow with great difficulty like very cold treacle. A consequence of the different behaviour is that basaltic flows tend to be relatively thin and very extensive, while silicic flows typically are short, stubby and rather thick.

Viscosity – which is linked closely to a magma's chemical composition – is fundamental in determining volcanic form. Resistance to flow – which is the definition of viscosity – is also a function of the proportion of volatiles the magma contains, and of its temperature. As a body of molten rock rises higher in the crust and vesiculation takes place, its viscosity

The steep-sided cone of Mount Teide, Tenerife. This stratovolcano is built from recent basalt, andesite and dacite flows and pyroclastic rocks. The steepness of the cone is largely a function of the predominance of fragmental rocks and silicic lavas within the sequence.

Steep-sided dacitic plugs intruded through the older shield basalts of Isla La Gomera, Canary Islands. In the foreground is Roque de Agando.

The low-profile basaltic shield volcano, Nea Kameni, Thira, Greece. This young volcano has been growing since the sixteenth century, from the floor of the Aegean Sea within the ancient caldera of Thira, which exploded between 1400 and 1500 BC.

may rise dramatically; the result is that many already rather viscous silica-rich magmas tend to clog up their vents, erupting explosively as the pent-up volatiles dismember the magma, generating massive explosion craters and huge ash clouds that cover wide areas. Mounts Pinatubo and St Helens are two cases in point. Even low-silica basalts generate fire-fountains, but these tend to be very localized, consist of molten lava droplets and do not spread pyroclastic debris over large areas.

Temperature also affects viscosity. Thus, in time, all lavas begin to cool, causing the viscosity to increase until a point is reached when the forward movement of the flow is prevented by the resistance of the cooling lava to further movement. Flow then ceases and crystals rapidly begin to form in the magma. When all of the liquid has been converted to crystals by the crystallization process, an igneous rock has been formed. When a large volume of fluid magma issues from a vent or fissure at a high rate, flow of the fluid lava will continue for a long period, with the result that extensive regions may be covered; such activity produces what geologists term flood lavas. This style of activity accompanies the splitting apart of continents, manifesting itself in vast plateaux built from basaltic flows, such as occur in the Deccan of India, the Columbia River region of the Western USA and the Hebridean region of Great Britain. More modest rates of eruption tend to build shield volcanoes like those in Hawaii which are some of the largest volcanic structures on the Earth. Despite their large size – typical Hawaiin shields measure several hundred kilometres across their bases – they do not make impressive landscape features since their flank slopes are usually less than one degree.

Much more impressive are the steeper-sided volcanoes built by more silicic lava and their associated pyroclastic deposits. Good examples are East Africa's Kilimanjaro and its near neighbour, Meru, Mount St Helens and Lassen peak in America's Cascade Ranges, and dangerous volcanoes such as Gunung Merapi and Gunung Agung in Indonesia. While not as huge as large shields, they make very spectacular features in the landscape, as well as presenting mankind with hazards that are difficult to predict and take steps against.

Eruptions that take place from a central vent generally produce symmetrical structures, like Vesuvius or Stromboli; eruptions from fissures or elongate vents, produce asymmetric landforms, for example those associated with Kilauea's rift zones, or the rift-related structures of the Snake River Plain, in Idaho. Some eruptions, particularly those which result from the interaction of hot lava with groundwater – called *phreatomagmatic* eruptions – achieve more destruction than construction, typically forming explosion craters, or *maars*. Fine examples of these may be seen in the Auvergne region of France and in the Eifel area of Germany.

The obvious relationship that has been established between the properties of terrestrial magmas and the landforms that are produced under differing conditions, allows planetary geologists to use the information to make comparisons of them with their counterparts on other worlds. Comparative studies have proved highly informative in the quest to understand the geological histories of these neighbours of ours, as well as widening our knowledge of how magma behaves under different gravity, atmospheric pressure, and surface temperature. One thing which is sure, is that volcanism – in one form or another – has been one of the major geological activities on most of the solid worlds so far visited. It is a natural spin-off from the activity of a planet's heat engine. The Earth's heat engine appears to have remained more vigorous than most, driving a range of geological processes that continue to the present day and are likely to do so for many hundreds of millions of years yet.

6

Magma

What is magma?

Photographs or films of molten lava escaping from a volcanic vent or fissure are now commonplace; most people will have seen them at one time or another. The issuing of lava in this way is the surface expression of volcanism, a process which is instigated within our planet – and others – by a process called partial melting. Before explaining how this operates, let us begin by answering the simple question: 'What is magma?'

Rising magma

The rise of magma. As hot fluid magma rises, it cools. Crystals usually form during uprise, while the contained gases expand as the pressure diminishes, forming bubbles. These expand as the pressure further decreases, aiding the upward movement.

Magma is simply molten rock. It is the material from which all igneous rocks form by the process of crystallization. Ignoring the details of exactly how magma originates for a moment, any body of magma will be at high temperature and pressure when it resides within the Earth's crust or upper mantle; however, as it moves towards the surface it becomes cooler since it loses heat by conduction to the rocks through which it passes, and eventually it will cool sufficiently for crystals to form within it. At the same time, if it is rising towards the surface, the amount of rock above it will gradually become less, with the result that the load pressure exerted on the magma also diminishes. Prior to the onset of crystallization the atoms of the chemical elements it contains will have no ordered arrangement, moving around relatively freely within the liquid. As soon as the freezing temperature of the more refractory minerals that can crystallize from such a melt is reached, the atoms begin to move much more sluggishly and eventually will be unable to move at all. At this point crystallization has begun and parts of the system take on the ordered crystalline state typical of all solids.

Insofar as we are aware all of the magma which develops on the Earth, the Moon and other terrestrial planets, contains silica; i.e. silicon combined with oxygen, e.g. SiO_2, SiO_4 etc. Combined with the silica are other elements which occur in varying but significant amounts; such *major elements* include aluminium, titanium, iron, magnesium, calcium, sodium, potassium, phosphorous and manganese. Generally speaking the proportion of silica ranges between 35 and 75 per cent, depending upon the type of magma; basalts typically contain between 40 per cent and 48 per cent SiO_2, while granites hold between 68 per cent and 75 per cent SiO_2. In addition to the major elements, there are others whose amounts are measurable, not in per cent, but typically in parts per million; these are termed *trace elements*. Nickel, cobalt, rubidium and zinc are typical elements of this kind.

In addition there are also significant amounts of volatiles that are dissolved as gases within the magma. Water is the most important of these, but it is accompanied by such elements as boron, chlorine, fluorine and compounds of sulphur, hydrogen, oxygen and nitrogen.

While the magma stays deep inside the Earth, these more mobile elements and compounds remain dissolved within it, but as it rises towards the surface and the load pressure reduces, the volatile components may be released as gases such as carbon dioxide and hydrogen sulphide. Consequently, when a magma eventually reaches the surface likely it will consist of a mixture of molten silicate, crystals and gases.

Inside the Earth

Despite stories such as '*Journey to the Centre of the Earth*', there is not the remotest chance that humans will ever be able to descend to the core of our planet and directly see what is there. What we have learned about the inside of our world has been gleaned largely by geophysics, in particular the study of earthquakes. These are generated when rocks slip suddenly against one another, usually along fractures called *faults*. When this occurs three kinds of seismic waves are produced: compressional or *P-waves* which move in a push–pull manner; shear or *S-waves* which vibrate back-and-forth in a direction perpendicular to their travel path, and longitudinal or *L-waves* which travel along the Earth's surface. Both P- and S-waves can be transmitted freely through solids, but only S-waves can pass through liquids. This behaviour has been of prime importance in establishing the internal structure of our planet.

Seismic waves behave rather like light waves in that they can be reflected from, transmitted through or refracted by layers of rock within the Earth. The denser the medium through which they pass, the

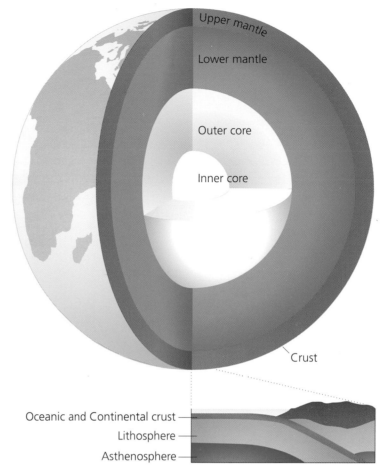

Density kg m^{-3}	Pressure kbar	Temp °C
3000		1000
4000	250	1850
4500		
5500	1400	3000
10,000		
12,000	3400	3700
12,500		
12,500	3850	4000

Inside the Earth. Diagram showing the internal layering and the way in which temperature, pressure and density increase with depth. Sudden changes are accompanied by abrupt alterations in the velocity of seismic waves – 'seismic discontinuities'.

higher is their velocity – a useful fact when considering the density distribution below the Earth's surface. Furthermore, when the waves encounter a sharp change in density, they become refracted, just as light rays do when passing through a lens or prism. By analysing the paths and velocities of seismic waves, geophysicists have been able to build up a pretty accurate picture of the Earth's internal stratification.

Natural seismic waves take a variety of different paths through the interior. For a start, the foci of earthquakes are found to be at varying depths, thus the waves generated by a seismic event will need different amounts of time to reach the surface. By mounting a sensitive instrument called a *seismometer* – or to be more precise, a series of such instruments – at the surface, it is possible to establish the depth of the earthquake focus and the density of the rocks through which the associated waves have passed. A high degree of accuracy is achieved by setting up a

network of very sensitive seismometers all over the world, and studying the pattern of the various types of waves on the *seismographs* produced. It has been established that the outer boundary of the Earth's core strongly reflects both P- and S-waves; in fact the S-waves cannot pass through it. This shows that the outer region of the core must be liquid. Indeed, it is within this liquid outer core that the Earth's magnetic field is generated.

Seismic discontinuities can be picked out on seisograph records quite clearly. In particular, the outermost 70 km of the Earth can be separated from the underlying layers, there being a sharp increase in the velocity of the waves below this level, a level marked by the Mohorovicic Discontinuity. This boundary defines the base of the *lithosphere*. Above this lies a part of the Earth which we can directly sample, namely the oceanic and continental *crust*. The lithosphere is characterized by strength and solidity.

The underlying layer is relatively weak and seismic waves are attenuated as they pass through it. It has become known as the Low Velocity Zone (LVZ for short) and marks the top of the *asthenosphere*. This region appears to be a stiff mixture of crystals and magma which, chemically, is rather like the lithosphere and is believed to be composed of a rock called peridotite, made up largely from the mineral olivine ($[MgFe]_2SiO_4$). With increasing depth the velocity of seismic waves again rises, the rocks becoming denser and denser. Suddenly, at a depth of 400 km, there is a sharp increase in wave velocity. This puzzled geologists for some time, but experiments involving the study of olivine at varying temperatures and pressures, revealed that it undergoes a *phase change* when the pressure attains a critical value. It seems that the 'normal' form of olivine, if subjected to very high pressures, changes to a more compact form in which the individual atoms are pressed closer together; the

olivine changes to another phase, namely spinel. It is this phase transition which marks the base of the asthenosphere.

Thus, we find that the Earth's *mantle* can be subdivided into a series of layers whose boundaries manifest themselves in abrupt changes in the behaviour of seismic waves. At greater depths than the base of the asthenosphere there are further discontinuities, but the next really major one occurs at a depth of about 2900 km, where the mantle meets the *core*. This, as we have already seen, is liquid in its outer part but solid again at the centre of the Earth.

Where and how do magmas form?
That magmas exist is beyond dispute. Not so is the topic of exactly how and where they form – this has engendered much discussion during recent years. One very important observation that can be made is

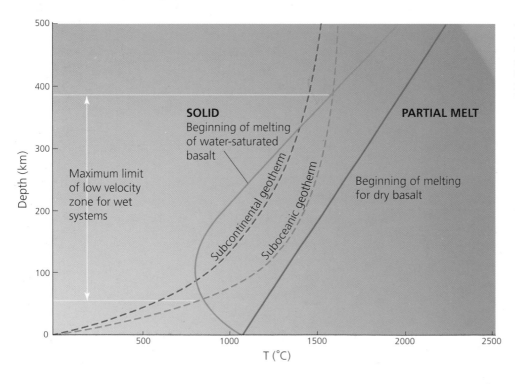

Thermal gradients in the upper mantle, showing the wet and dry melting curves for peridotite. Partial melting of peridotitic mantle produces basalt under upper mantle conditions.

Diagram showing the formation of diapirs below the level of the Moho and the probable arrangement of magma bodies beneath a typical mid-ocean ridge.

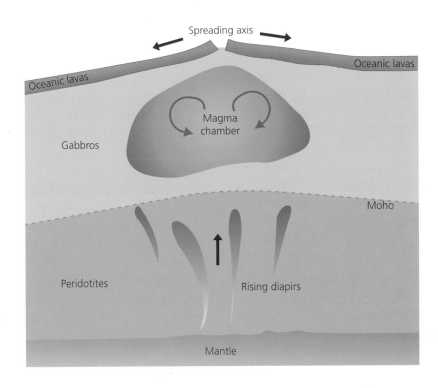

Granite outcrops at Granite Dells, Prescott, Arizona. Silicic rocks such as granite often weather into bizarre rounded outcrops as weathering exploits the well-jointed material. Beyond the Dells can be seen more sharply-defined outcrops of dark basalt.

of where they emerge at the Earth's surface. The most active sites – regrettably the least accessible to human view – are to be found along what are termed *mid-oceanic ridges*. These are global-scale submarine mountain ranges, very linear and quite narrow, usually with rift faults along their axes. The rifted crests are also the sites of volcanic vents and fissures from which issue large volumes of fluid basaltic magma and sulphurous gases. Active fumaroles and lava vents have been observed from submersibles, such as *Alvin*. The bulk of present-day basalts appear to emanate from such sites.

Basaltic lavas also are found on oceanic islands and on the Earth's continents; however, silicic magmas such as dacites and rhyolites are more typical of continental regions and also of certain island regions, particularly along the margins of the Pacific Ocean, where andesites are particularly common. Most volcanically-active zones tend to be distinctly linear; this is true of oceanic ridges and also of island arcs, and indeed of continental volcanic zones which tend to hug the continental margins. The linearity of volcano distribution is related to *plate tectonics*, a process whereby major sutures in the Earth's lithosphere allow movements between adjacent slabs of the crust, and also concentrate geological activity within narrow zones. Along with volcanicity, earthquakes occur, while the formation of mountain ranges also has its beginnings here; furthermore, it is the underlying reason why different types of magma characterize continental and oceanic regions.

To appreciate why magmas vary from place to place, it is necessary to investigate how they form in the first place. When they crystallize, magmas consist of a variety of silicate minerals, each of which has its own specific freezing temperature/melting point. Typically these range between about 750 and 1200 °C. If a magma contains substantial amounts of water, these temperatures are lowered. The presence of water (and other volatiles) thus allows rocks to melt at lower temperatures than they otherwise might if they were 'drier'. Most magmas contain some water (the Moon's are an exception) and that volatiles are common is seen in the omnipresence of volatile-filled veins and cavities in the vicinity of magmatic bodies, even small ones. Some of this volatile material will have originated in the magma itself, but a lot may have been contributed by the adjacent rocks, in which groundwater was circulating at the time the magma was injected, or pore-water trapped.

Temperatures and pressures in the upper crust are generally too low for solid silicate rocks to be melted; at greater depths, however – say around 70–100 km – we might expect at least some melting, even of 'dry' rocks. At such depths we would be dealing with upper mantle material under considerable pressure and at elevated temperature. The prediction that some melting should occur in this situation is borne out by seismic data, which indicates that a partially molten zone exists here. It is within this Low Velocity Zone of the asthenosphere that geologists believe most basaltic magmas are produced, by selective melting (known as *partial melting*) of some components of the upper mantle materials. More melting appears to occur than theory predicts for 'dry mantle', and it has to be concluded that volatile material somehow enters the equation.

The presence of volatile components in the lattices of hydrated minerals such as mica and amphibole – both of which have been found in exotic blocks brought to the surface from mantle depths by some violently explosive eruptions – may have helped to catalyse melting at these depths. Additionally, deep-seated fracturing might also aid in reducing pressures below critical levels. We also know that the process of plate tectonics involves the constant recycling of the crust, together with the water-laden sediments which become deposited upon it; where the water-laden sediments are taken down inside the Earth, this provides the volatile material necessary to reduce the melting temperatures of the solid rocks below. Melting may therefore take place at shallower levels than it would if the mantle were 'dry'.

The origin of the hydrated, silica-rich magmas of the continents may be rather different. Modern theories generally appeal to the idea that continental-type magmas are, at least in part, produced by partial melting of a variety of water-rich crustal rocks, i.e. quartz sandstones and existing igneous rocks. Some kinds of magmas, particularly andesites (moderate silica content) which are typical of active belts that encircle the Pacific Ocean, may have their origins partly in the partial melting of mantle rocks, and partly in that of silica-rich sediments or igneous rocks of similar composition. In the presence of substantial amounts of volatiles such a mix could probably produce silica-rich magmas at depths as shallow as 35–40 km. As we shall see in Chapter 8, the ideal sites for such a process are encountered at what are termed subduction zones.

The critical factor in all of this is that for partial melting of solid rocks to occur, at the appropriate depth, there has to be a change in the temperature, pressure or chemical composition of the starting material if any fraction of it is to commence melting. Thus, if the mantle were unusually hot in one spot, perhaps due to it having been bequeathed an abnormally large inventory of radioactive elements, melting might begin. Then again, if the pressure were released along one of the crustal sutures we know to exist, again, melting might be instigated. Indeed, the requisite changes needed for melting to take place are most likely to occur along the boundaries between the slabs of lithosphere which make up the Earth's segmented outer shell.

When melting does occur, the first liquid fractions that develop will contain the least refractory components. (This is the opposite of what happens during crystallization, where the most refractory materials crystallize first.) As the temperature gradually

increases, so more refractory components will be added to the partial melt; and so on. Such pockets of melt will be less dense than the adjacent solid mantle and therefore they will tend to rise, usually as bulbous bodies termed *diapirs*. If, during their upward motion, some crystallization begins to take place (they will tend to cool as they rise higher in the crust), then the first crystals to form will be those of the more refractory silicates – for instance, olivine – which, being denser than the melt, will start to sink down to the bottom of the magma body. In this way the originally homogeneous pocket of melt will gradually undergo a chemical change, with the solidified portion being enriched in the relatively dense, refractory elements (e.g. magnesium, chromium and iron) and the overlying, still-liquid portion becoming relatively richer in less refractory elements, e.g. silica, potassium, aluminium and sodium. This process, called *fractional crystallization*, is fundamental in generating magmas of different composition from place to place within the Earth.

The lesson of Hawaii

Over a century ago the famous American geologist J. D. Dana noted that there was an increase in age along the Hawaiian chain of islands, from the active volcano of Hawaii in the southeast, to the heavily eroded island of Niihau to the northwest. A similar trend was noted for the Canary Islands in the Atlantic Ocean, and subsequently has been suggested for several other of the Pacific island chains. The Hawaiian volcanoes undoubtedly are the most closely studied on Earth, and so it seems logical to enquire what we have learned from all this intense investigation.

The islands form the southeastern end of a long series of atolls, reefs, submarine mountains and submerged volcanic cones that extends 6500 km across the Pacific floor. The volcanoes within this chain run stepwise rather than in a simple line across the ocean

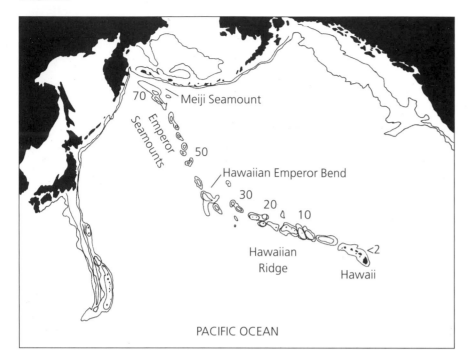

Ages of volcanoes of the Hawaiian–Emperor chain (millions of years). Currently volcanoes on the island of Hawaii itself are active; with increasing distance towards the northwest end of the chain, successively older centres are encountered. The bend in the chain appears to be a function of a change in the pattern of plate movement about 40 million years ago.

floor, the individual eruptive centres being spaced roughly 75 km apart. Most of the islands are built from two or more volcanoes, Hawaii itself being formed from five: Kilauea, Mauna Loa, Mauna Kea, Hualalai and Kohala. The greater volume of each is submerged beneath the water. Thus the oceanic rise upon which the chain sits is elevated about 600 km above the ocean floor, and then a further 4000 m above sea level on the island of Hawaii itself. The flanks of the volcanoes have gentle slopes, ranging between 5° and 10°, the steepness tending to increase below sea level.

It is the age pattern shown by the islands that is of extreme interest: there is a gradual increase in age from Hawaii in the southeast, where there is current activity at the centres of Kilauea and Mauna Loa, until we reach the 5.8 million-year-old centre of Kaui, some 520 km northwest. When age data for all members of the chain are considered, it is clear that activity has been ongoing for at least 65 million years. The youngest volcano has not yet built up to sea level: this is Loihi, which sits off the southeast coast of Hawaii.

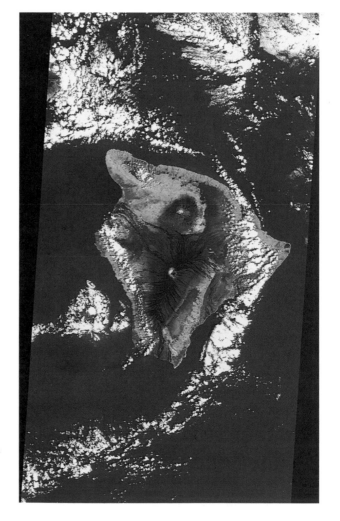

Satellite image of the Big Island of Hawaii. The brownish, snow-spattered cone occupying the northern part of the island is Mauna Kea. To the south lies Mauna Loa, with its distinct summit caldera. Dark basaltic flows can be seen spreading out from the southwest/northeast Rift Zone which crosses the mountain. The caldera and recent bluish-grey basalts of Kilauea lie adjacent to the southeast coast. [Photo: USGS Flagtsaff Image Facility]

Pahoehoe lava flooding the floor of Mokuaweoweo caldera, Mauna Loa, Hawaii. [Photo: Patrick Moore]

What appears to have happened during this period is that the focus of volcanic activity has slowly shifted southeastward, at a rate calculated to be around 8 cm per year. In 1963, J. Tuzo Wilson expounded his famous theory that the age variation of the chain developed as a result of the lithosphere in this region moving slowly but continuously northwestward over a fixed *hot spot* in the underlying mantle. He suggested that under the island of Hawaii was a currently active upwelling of hot mantle material – a *plume* – that has been in existence for many tens of millions of years, and has supplied magma to the surface more or less continually during this time. Volcanoes that formed over the magma source were successively moved northwestward, away from the heat source, eventually becoming extinct as the magma supply was cut off.

The hot spot idea has become widely accepted and modern estimates suggest that the Hawaiian one has a diameter of about 300 km and is located some 50 km northeast of Kilauea volcano. Other workers have extended Wilson's original idea, suggesting that hot spots are the result of upward movement of a plume of hot mantle material from great depth, this being driven upwards by convection. What better example could there be of the Earth's heat engine at work! Typically, hot spots express themselves at the Earth's surface in widespread volcanism, positive gravity anomalies and a higher than normal outflow of heat, while the mantle beneath tends to have unusual characteristics. Interestingly, the style of volcanism on Earth's sister planet, Venus, appears to be dominated by major plume activity, rather than by volcanism associated with plate tectonics.

Vesuvius and all that

The volcanoes of Hawaii reside completely within an oceanic region, and the active centres erupt every few years. The low-angle structures are called *shields* and are dominated by basaltic rocks. Things are very different elsewhere. For instance, the central part of the Mediterranean region lies between two relatively stable zones of the Earth's lithosphere that are moving inexorably towards one another. In consequence it is an area of intense geological activity. Active island volcanoes are to be found on Sicily (Etna), Stromboli, Vulcano, Nisyros, Nea Kameni and at one continental

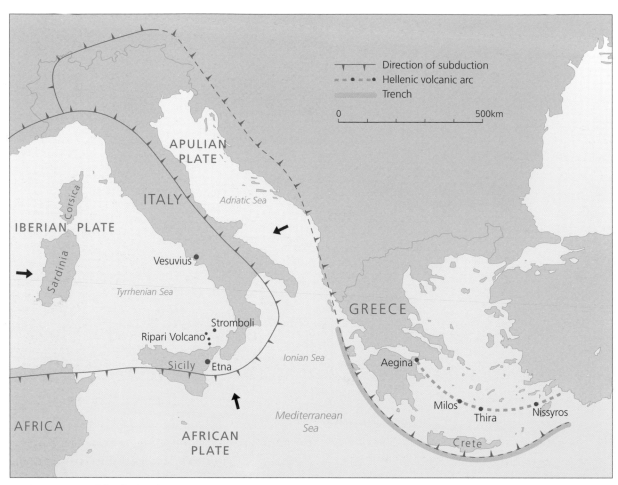

The plate structure and distribution of volcanoes in the Mediterranean region.

Mount Vesuvius viewed across the Bay of Naples. The recent cone (right) has built up within the older, partially destroyed, caldera of Monte Somma (left).

Pyroclastic surge deposits associated with the catastrophic eruption which overwhelmed the town of Herculaneum in AD 79. Photographed in Herculaneum.

site – Vesuvius, probably one of the most famous volcanoes in the world. Earthquake activity is a commonplace and mountain building a characteristic of this area. It is a very mobile zone.

Nearly all of the volcanoes in the Mediterranean are of the type termed *stratovolcanoes*; this means that they are constructed from a mix of lavas and *pyroclastic* rocks, i.e. fragmental rocks produced by explosive eruptions. In this they are very different from the Hawaiian shields. Their flanks are much steeper in profile than shields, with angles of between 20° and 30° common. Normally a central vent (a crater) sits atop the summit of a well-stratified cone.

Activity at Vesuvius commenced about 10 000 years ago, since which time at least four different volcanoes have occupied the site. The modern cone sits within a large depression encircled by the ruined walls of an older volcano, Monte Somma. Typically, Vesuvius – which lies a mere 10 km from the centre of the densely-populated Italian city of Naples – has an irregular cycle of eruptions with 25- to 30-year periods of quiescence punctuated by major outbursts. During one of the latter, a huge plume of ash and

dust will rise over the city, spreading many kilometres into the atmosphere. The most recent eruption, that of 1944, saw Allied airfields in the Naples area being carpeted in ash, forcing planes to be grounded for a short period.

The most devastating eruption was that recorded by Pliny the Younger in AD 79. During this massive outburst, huge volumes of ash, dust and blocks were hurled out, burying the heavily-populated towns of Pompeii and Stabiae nearby. Although located 8 km from the focus of the eruption, Pompeii was covered to a depth of 3 m in pumice. The neighbouring town of Herculaneum had an even more appalling fate, it was overrun by an incandescent, ground-hugging cloud of toxic gas and ash which also set off mudflows. During the two days of eruption, roughly 3 km^3 of material was ejected from the volcano, which then suffered collapse.

A high degree of explosivity characterizes Mediterranean volcanoes in general. This is because the magmas generated in such regions of the Earth's crust are more silicic than their Hawaiian counterparts, and consequently their lavas much more vis-

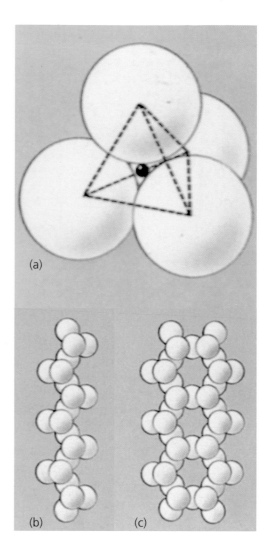

Atomic structures of silicate minerals: (a) basic silicate tetrahedron – four oxygen atoms surrounding one silicon atom; (b) single chain structure; (c) double chain structure.

cous. Dacite, rhyolite and andesite are far more common than basalt. The spectacular water-filled collapse caldera of Thira (Santorini) is one of the wonders of this region. The original stratovolcano underwent a massive eruption 3500 years ago, sending pumice and ash over a very wide area, setting off tidal waves and earthquakes and, according to some scientists, played a major part in the demise of the hitherto highly successful Minoan civilization. It then collapsed to form a huge depression which was filled by the sea.

Earthquakes are far more numerous and much more dangerous in these kinds of location than in oceanic areas. During seismic activity associated with Vesuvius, 2909 people were killed at Laviano in 1980, while in 1908, over 85 000 people were killed in the town of Messina, in northeastern Sicily, by quakes focussed beneath the island. The reason for the greater danger from earthquakes is that stratovolcanoes typically are associated with the boundaries between the Earth's lithospheric plates, where volcanism is nearly always accompanied by seismic activity, often of some severity. The earthquakes are triggered by sharp movements between two or more plates, where they grind against or slip past one another.

The silicate minerals

The two elements, silicon and oxygen, are by far the most abundant in the Earth's crust and in the mantle. Thus, when magma crystallizes and rocks form; these are composed largely of silicate minerals. The order in which specific minerals appear in the sequence of crystallization is determined by the temperature and pressure of the magma, and on the amount of volatiles it contains. Generally speaking, those minerals with the simpler structures crystallize first; later, as the magma cools, more complex silicates which tend to contain more volatiles crystallize.

A typical silicate mineral will be a combination of silicon and oxygen with one or more metallic elements. Such silicates show a wide range in chemical composition, external appearance and physical properties, but each is based on an internal structure called the silicate tetrahedron. Silicon and oxygen naturally combine in the ratio of one silicon atom to four oxygens (SiO_4), the oxygens being positioned around the solitary silicon atom so that they occupy the four apices of a three-dimensional figure called a tetrahedron. This SiO_4 'anion' is extremely versatile since it can form complex chains and sheets of tetrahedra, making giant molecules. In these, some or all of the oxygens belonging to individual tetrahedra are shared with adjacent ones; the precise way in which they are arranged allows them to be classified into a small number of silicate mineral families. Within these families the chains, rings or sheets of silicate

(a)

PHOTO 24 - (a): Purple fluorite crystals embedded in milky calcite. (b): Crystals of white barite. (c): Rounded crystals of the iron oxide, haematite.

anions are linked together by a variety of metal 'cations', such as iron and calcium.

In the simpler families the individual tetrahedra remain rather as islands, sharing none of their oxygens and being joined together by cations like iron and magnesium. Olivine has this kind of structure. With the sharing of more and more oxygens between adjacent tetrahedra, larger holes develop in the crystal lattices and larger cations, like calcium, sodium and potassium, can be accommodated (e.g. pyroxenes). Indeed, in silicates belonging to the amphibole and mica families, even the large hydroxyl (OH) molecule can be fitted in, giving silicates which are less dense than the simpler, compact ones such as the olivines and pyroxenes.

Because the electrostatic bonding between silicon and oxygen atoms of the silicate tetrahedra is substantially stronger than that between the tetrahedra and other metallic cations, crystal morphology and lines of weakness (cleavages) within their crystals are predominantly a function of the SiO_4 tetrahedral arrangement. The various crystal frameworks are shown in the accompanying diagrams.

(b)

The non-silicate minerals

Not all of the minerals associated with magmas are silicates. Within most bodies of igneous rock there are small amounts of non-silicates, like oxides, carbonates and sulphides. Exceptionally some oxides, for instance, will accumulate as layers of crystals within a magma chamber, producing important ore deposits – the chromite deposits of the famous Bushveld Intrusion of South Africa are a case in point. Most of the non-silicates, however, appear to be carried away from their parent magmas by volatile-rich fluids which may later deposit them elsewhere, effecting what is termed 'mineralization'.

Metals such as iron, magnesium, lead, copper and zinc may eventually be concentrated sufficiently to

(c)

form ore deposits – non-silicate bodies of sufficient volume and concentration to be economically viable. Most of these leave the parent magma in the form of metal chlorides which are held in solution as hot aqueous fluids. These fluids will gradually escape towards the Earth's surface along fractures or through the pores or weaknesses in permeable rocks. As they travel through successively cooler rocks they will lose heat, their temperature falling such that they enter into chemical reactions with various kinds of rock through which they pass. At successive stages in their upward passage, different metalliferous minerals are deposited, usually in order of diminishing melting point. Typically the sequence is oxides → chlorides → sulphides → carbonates, the element sequence being tin → copper → zinc → lead.

The deposition of ore minerals and of the non-metallic minerals called gangue minerals which accompany them – of which quartz is by far the most abundant – is a very lengthy process by everyday standards. For a small body, deposition will take decades, while a large one will need millenia to form. However long it takes, the metal-bearing fluids gradually cool with time, the result being that even very close to the magma source, temperatures will have dropped to several hundred degrees below their initial values. This being so, the high-temperature minerals precipitated near the source during the early stages of mineralization will be joined much later by lower-temperature minerals, giving rise to what is termed a *zoned* ore body. The zonation produced as a result of the original temperature gradient is termed *space zoning*, while the zonation generated later, as the temperature falls, is called *time zoning*.

The complexities of ore deposition are considerable but, because humans require metals to carry out what have become everyday operations, they have to be constantly researched to optimize the chances of finding and exploiting viable deposits. In recent times satellite imagery has been used to search for new deposits, while other sources have been discovered by using submersibles to explore the ocean floor.

How hot is it inside the Earth?

Measurements made by delicate instruments lowered into deep mines and bore holes indicate that the average rise in temperature per 100 m of depth, is between 2 and 3 °C. Temperature gradients can only be measured in this way down to depths of around 8 km, however, beyond this other methods have to be employed.

Geophysicists have to use sophisticated means to discover how much heat is flowing out from the Earth, but in recent years they have been able to measure, at least for the continental crust, what the surface heat flow is. Having done this, calculations can be made which reveal that in tectonically-active parts of the continental landmasses, temperatures of the order of 1000 °C are encountered at depths of around 40 km. In contrast, at the same depth under more stable continental regions, the geothermal gradient is less and the temperature rises only to about 500 °C. Significantly, volcanism and crustal mobility are associated with regions of high heat flow, while geological stability is associated with those of lower heat flow. Under the oceans the outflow of heat is greatest along oceanic ridges, where seismic activity and volcanism are concentrated. Elsewhere the temperature on the ocean floor is probably close to 0 °C, rising to about 1200 °C at the base of the lithosphere.

At present there is no direct way of telling what temperatures characterize greater depths, but the indications are that at the centre of the Earth the temperature approaches 4300 °C, while at the core–mantle boundary, a figure of approximately 3700 °C is likely. Finding a way of making more accurate determinations at these great depths is just one of the as yet unsolved problems for geologists.

7

The perpetual dynamo

Magnetism

Lodestone, with its power to attract or repel pieces of iron, has been known since ancient times. It is in fact an oxide of iron called magnetite. The Chinese seem to have known 3000 years ago that a magnet will point toward the North Pole when freely suspended and allowed to swing in any direction. Yet this escaped the Western world until the twelfth century, when it was first used in that invaluable navigational aid, the mariner's compass.

In the earliest form, the water compass, a piece of magnetized iron was placed in a wooden vessel and floated on water, so that it could move freely in a horizontal direction. Later mariners used a pivoted mag-

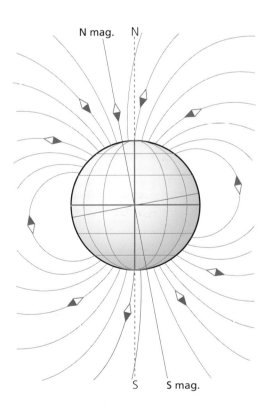

The Earth's magnetic field is best considered as being produced by a huge bar magnet at its centre. The alignment of the magnet does not quite coincide with that of the Earth's spin-axis. A suspended magnetized needle would align itself in the direction of the magnetic field as shown.

netic needle with a disk surmounting it. This was known as a compass card, and was divided into 32 equal 'points'.

In 1269 Picard Petrus Peregrinus described how he located the magnetic poles of a lodestone globe, and also showed how like poles repel one another while unlike poles attract. Famous mariners such as Columbus and Magellan used compasses, but the first really major advance was made by the English physicist William Gilbert. His book *De Magnete, Magneticisque Corporibus*, published in 1600, discussed magnetic variation and contained the suggestion that the Earth itself behaves like a giant magnet.

Gilbert could not explain how a compass worked, but in the early nineteenth century experiments with simple batteries showed how magnetic forces are generated. Magnetism was, in fact, a function of electrical forces. These should really be termed *electrodynamic* forces, but for historical reasons they are normally known as magnetic forces.

The Earth's magnetic field

As Gilbert suggested, the Earth behaves like a huge magnet; if a magnetized needle is suspended freely, it will align itself in the direction of the local magnetic field. The easiest way to represent a magnetic field is by lines of force; that is to say, by a series of lines everywhere pointing in the direction of the field. Where the field is strong, the lines of force will be

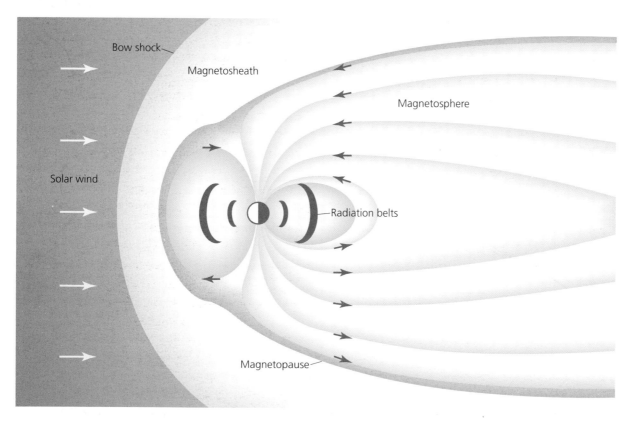

The magnetosphere is roughly teardrop-shaped, with its tail pointing away from the Sun. The asymmetry is created by interaction between the field and the high-energy particles of the Solar Wind. Where the two meet a shock wave is produced (bow shock), within which is a turbulent zone bounded on the inner side by the magnetopause. The magnetosphere proper lies on the Earthward side of the latter.

crowded together; with a weak field the lines are more widely separated. The lines are, therefore, contours of the strength of the field.

The Earth's field takes the form of a magnetic dipole, behaving the same way as the field produced by a bar magnet. Indeed, the best way to understand the features of the Earth's field is to picture a huge bar magnet running through the centre of the Earth itself. To be more specific, the axis of the magnet is almost parallel to the geographical north–south line or spin-axis; at present the angular difference is about 18°. The north magnetic pole is located in the Arctic islands of Canada; the south magnetic pole lies south of Tasmania, although, as will be seen in Chapter 13, this has not always been so; the positions of the poles are not constant.

The field strength at any point on the Earth's surface is measured by a delicate instrument known as a magnetometer. The unit of field strength is the *gauss*, named after the famous nineteenth-century German mathematician J. K. F. Gauss. At the Equator, the field strength is 0.3 gauss; at the magnetic poles it is roughly double this. To put this in perspective, it is worth noting that a typical horseshoe magnet has a field strength of about 10 gauss.

Although the picture of a huge bar magnet inside the Earth is a good analogy, it is an over-simplification, and it cannot be said that as yet we fully understand the real situation. However, this analogy is adequate for our present purpose.

The Earth's magnetosphere

The magnetic lines of force act as a shield round our planet, protecting us from harmful radiations from space. Moreover, high-energy, high-velocity particles constantly strike the upper atmosphere. There are, in particular, cosmic rays, which are not really rays at all but are atomic nuclei; some of them come from the Sun and from the planet Jupiter, although most origi-

nate far beyond the Solar System. When the cosmic ray particles strike the upper air, the heavy nuclei break up and only harmless 'secondary' particles reach the ground.

To check the cosmic ray counts at various altitudes, the American artificial satellite Explorer 1 was launched in 1958. This was in fact the first successful American satellite, and proved to be more important scientifically than the early Russian Sputniks. Explorer 1 entered an elliptical orbit with a perigee (minimum distance from Earth) at 350 km, and its geiger counter measured the cosmic ray intensities at different heights. The results were surprising. Each time Explorer reached a height of 950 km, the readings dropped to zero. It was then found that this was due to radiation so intense that the geiger counter had become saturated and put out of action. Later investigations showed that there are zones of intense radiation surrounding the Earth; these are now known as the Van Allen radiation belts, after the American scientist James Van Allen, who had been responsible for the equipment on Explorer 1.

The Van Allen belts are contained in the magnetosphere, which is really an extension of the magnetic force field into a doughnut-shaped region surrounding the Earth. On the Sun-facing side it extends outward for about 10 Earth radii, but on the side away from the Sun it streams out into a 'tail' for several millions of kilometres before losing its identity in the overall magnetic field that pervades the Solar System.

The asymmetry is due to the Solar Wind, a stream of ionized particles (plasma) pouring out from the Sun in all directions. Upon striking the Earth's force field the particles are decelerated, producing a 'bow shock' inside which there is a turbulent region known as the *magnetosheath*, where the pressures from the Solar Wind and the Earth's field are equal. Inside this again is an interface known as the *magnetopause*, which encases the magnetosphere proper.

Changes in the magnetic field

A magnetized needle taken to various points on the Earth's surface and allowed to swing freely will settle down at quite different angles to the horizontal. The dip shown by the needle is termed the magnetic inclination. Moreover, the needle aligns itself not with the geographical pole, but with the magnetic pole. The angular difference between the two is termed the magnetic declination. As early as the sixteenth century it was found that the position of the magnetic pole changes with time; this is termed the secular variation. Thus at the end of the fifteenth century the magnetic variation was 11.25°, but fell to only 6° by the end of the sixteenth century. Fairly good records exist for the past 400 years. If we retain the picture of a huge bar magnet inside the Earth, to achieve these kinds of changes this magnet must 'wobble'. Only a liquid core could move fast enough to produce changes of this order, and it is now generally accepted that magnetic forces do indeed originate in the Earth's outer core.

Rocks, minerals and magnetism

Iron is an ingredient of all natural magnets, but not all magnetic substances behave in the same way. Thus metallic iron (Fe) is said to be ferromagnetic, because all its tiny atomic magnets are aligned in the same direction, while minerals such as magnetite (Fe_3O_4) are ferrimagnetic, because some of their atomic magnets point in opposite directions. If a ferrimagnetic mineral is heated above a certain well-defined temper-

Termite mounds in the Australian desert near Darwin, Northern Territory. These contain tiny grains of magnetite which have orientated themselves in the direction of the magnetic pole.

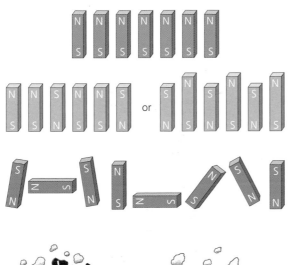

The atoms of iron in some minerals behave as tiny magnets. In ferromagnetic substances, like native iron, all the minute magnets share a common alignment (top). Ferrimagnetic minerals, however, have some atomic magnets pointing in one direction and some in the reverse, with an overall balance in one favoured direction (middle). If a ferrimagnetic mineral is heated above its Curie Point, it becomes paramagnetic, the common alignment being destroyed (bottom).

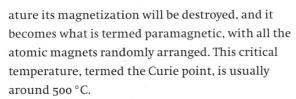

Magnetic and non-magnetic grains sink

Non-magnetic grains sink

Remanent magnetization can be preserved in sedimentary rocks. In case A (left) magnetic grains settling through water will be deposited on the seafloor such that their internal magnets become aligned in the direction of the prevailing magnetic field. In case B (bottom) magnetic material is being precipitated in the pore spaces between detrital grains and also takes on the prevalent magnetic field direction.

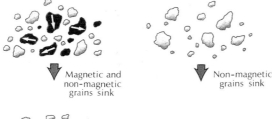

ature its magnetization will be destroyed, and it becomes what is termed paramagnetic, with all the atomic magnets randomly arranged. This critical temperature, termed the Curie point, is usually around 500 °C.

Exactly how a rock becomes magnetized depends upon the way in which it was formed. Rapid cooling, as with hot lava extruded on the Earth's surface, leads to crystallization; the temperature will fall below the Curie point, and the atomic magnets will align themsleves in the direction of the contemporary magnetic field (thermoremanent magnetization, or TRM). Other rocks may never become so hot, and will form by the settling of crystal fragments through water to form sedimentary rocks; if there are some already-magnetized grains in the deposit, they will tend to become aligned with the Earth's magnetic field (detrital remanent magnetization, or DRM). The Earth cannot be magnetized to great depths, because the temperature will be above the normal Curie point – which again destroys the simple picture of a bar magnet passing right through the globe.

Palaeomagnetism

Palaeomagnetism is the study of the fossil magnetization of rocks of all ages. It has been found that rocks containing magnetic minerals will bear the imprint of ancient magnetization, and measurements have shown that the magnetic poles have not always been in their present position: a phenomenon termed polar wandering. It has been deduced from this fossil magnetism that not only have the continents moved relative to the magnetic poles but also relative to one another. Another important discovery is that the Earth's field has not always had the same polarity; at various times in the past a compass needle would have pointed south, not north. There have apparently been at least nine such reversals during the past 4 million years.

Magnetometers carried on board ships or aircraft have detected magnetic anomalies beneath the ocean floor. A positive anomaly indicates that the rocks have the same polarity as the present geomagnetic field; a negative anomaly indicates a reversed polarity. These anomalies alternate, and trace out

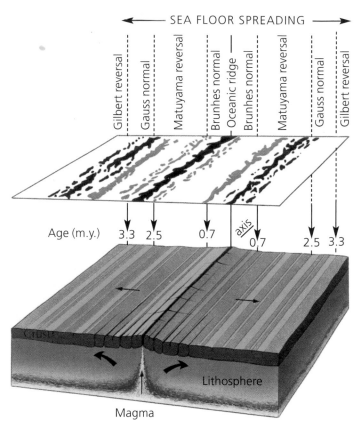

SEA FLOOR SPREADING

Gilbert reversal
Gauss normal
Matuyama reversal
Brunhes normal
Oceanic ridge
Brunhes normal
Matuyama reversal
Gauss normal
Gilbert reversal

Age (m.y.)   3.3   2.5   0.7   axis   0.7   2.5   3.3

Crust

Lithosphere

Magma

Polarity reversals are revealed by measuring magnetic anomalies in the basaltic rocks of the ocean floor. The alternating positive (green) and negative (brown) anomalies shown in the lower part of the diagram are related to linear belts of magnetized rocks on the sea-floor which were produced at the site of a spreading axis, and then carried away from it on both flanks by convection motions within the mantle. Lava extruded during periods when normal polarity prevailed gives positive anomalies, while the reverse is true for negative anomalies.

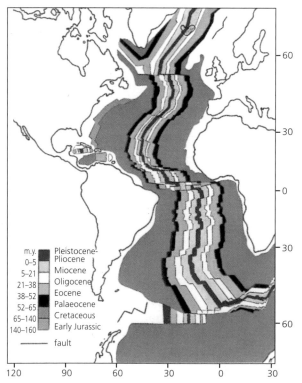

m.y.
0–5   Pleistocene–Pliocene
5–21   Miocene
21–38   Oligocene
38–52   Eocene
52–65   Palaeocene
65–140   Cretaceous
140–160 Early Jurassic
——— fault

A chart of the Atlantic sea-floor showing the pattern of magnetic striping. This reveals the symmetrical arrangement of the oceanic crust on either side of the Mid-Atlantic Ridge.

magnetic 'stripes' on the ocean floors, giving rise to patterns that extend for hundreds of kilometres.

We also known that submarine mountain chains rise from the ocean depths; these are called oceanic ridges. One of these – the Mid-Atlantic Ridge – divides the Atlantic Ocean along a roughly north–south axis. To either side of its crest the magnetic stripes are almost perfectly symmetrical, and similar patterns are found for all the other known major oceanic ridges.

It seems that the basaltic rocks of the oceanic crust act like natural tape recorders, recording successive periods of normal and reversed polarity when the relevant rocks poured out onto the ocean floor as lavas. We may assume, therefore, that oceanic ridge crests are linear lava conduits through which hot basaltic magma rises and flows out onto the sea-floor. The lava, when it cools, will fossilize the contemporary magnetization. Once this has happened, the polarity will not change; however, in time the lava is moved away from its source as the sea-floor 'spreads'. Radiometric dating shows that the strips of basalt flanking an oceanic ridge are progressively older with increasing distance from the ridge crest.

We can therefore determine the times of magnetic reversals. For example, a basalt extruded from the Mid-Atlantic Ridge about 10.5 million years ago will show reversed polarity, so that the Earth's field must have had reversed polarity at that time. This supports the theory of sea-floor spreading. For the Atlantic the present rate of spreading is about 1 cm per year in each direction away from the ridge crest.

The perpetual dynamo

The generally accepted theory of the Earth's magnetic field is due largely to the work of Sir Edward Bullard and W. M. Elsasser. It hinges primarily on the probability that the outer part of the Earth's core is composed of liquid iron and nickel. This means that it can conduct electrical currents. If the fluid is in motion, it can also interact with a magnetic field, and the field itself can influence movements in its interior.

Weak magnetic fields exist in the Solar System, and indeed all over the Galaxy. If for some reason or other the Earth's core were in motion, even a weak field of this kind could influence its movements. If the pattern of the movement happened to be just right, then a magnetic field could be created within the core. It is now generally believed that our planet had no magnetic field at all in its earliest period but that a field was subsequently developed by an interaction of this type.

The Earth has a fairly rapid axial rotation. This spinning motion has a profound effect upon fluid motions in the core, and results in the Earth behaving in the manner of a huge dynamo. Technically it is said to be a 'self-exciting dynamo', since once it began

to work it gathered momentum, regenerated the weak galactic field, and produced the relatively strong field of today. The exact mechanism is very complex, and is not yet fully understood; note, moreover, that there are suspicions that around the time of polarity reversals the overall magnetic field of the Earth may become very weak indeed, perhaps vanishing entirely for a time – which could have disatsrous effects upon life, since the protective screen would temporarily be withdrawn. However, the fact that the magnetic poles are not very far from the geographical poles (and, therefore, the rotational axis of the Earth) lends support to the idea of a dynamo process which is sustained largely by the Earth's axial spin.

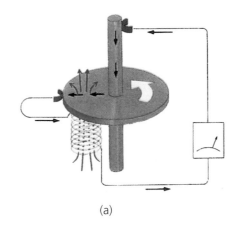

(a)

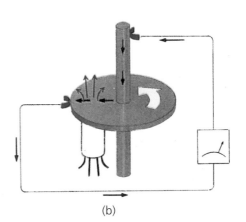

(b)

A simple diagram showing how an electric current is generated when a copper disc is rotated within a magnetic field produced by a bar magnet (a). If a coil is substituted for the magnet (b) the same electrical current produces a magnetic field which perpetuates the system. The terrestrial field is believed to operate in a similar, if more complex, way.

8

Continents and oceans

What is the lithosphere?

The Earth has a layered internal structure, consisting of a core, mantle and crust. Part of the core is believed to be liquid, the mantle, by-and-large is solid but behaves in a plastic fashion, while the crust is rigid. Historically, this tripartite division of the layers was sufficient to describe our planet's structure. However, in recent years it has become apparent that the outer layer is a little more complex than this.

Geophysical research has revealed that the brittle outer crust and the uppermost part of the mantle are coupled together and respond as a single unit with respect to movements in the mantle below this level. Together these layers have become known as the *lithosphere*, the principal characteristic of which – rigidity – is what preserves the topography at the Earth's surface. Immediately beneath this is a layer in which seismic waves are severely attenuated; this is termed the Low Velocity Zone, or *asthenosphere*; it appears to end about 200 km below the surface. It is at this level that most magmas are believed to form, and it is on this that the lithosphere 'floats', responding to whatever motions are taking place there. The thickness of the lithosphere is not entirely certain, but probably varies between 60 and 100 km.

Oceanic crust

The oceanic crust also sits lower than the continental crust, by between 3 and 4 km, and has a higher density. The average density is between 3.0 and 3.1 g/cm^3. At the present time there is no oceanic crust older than about 200 million years. The ocean floors are veneered with sediments, immediately beneath which are basaltic rocks forming a layer between 0.5 and 2.5 km thick. These in turn are underlain by about 5 km of coarser gabbros, which are separated from the asthenospheric mantle by a rather thin layer of olivine-bearing igneous rocks of somewhat greater density; the latter are called peridotites.

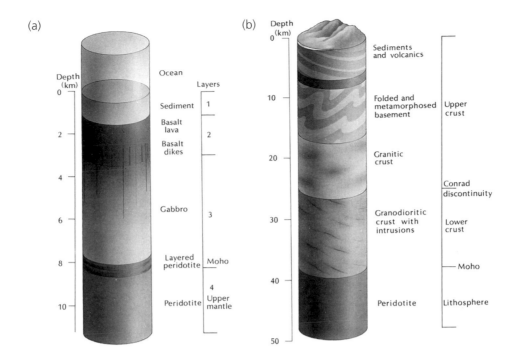

(a) A section through the oceanic crust, based on seismic data, dredgd samples and interpretation of other geological data. The relatively thin layer of recent sediments is underlain by layers of denser basaltic rocks, which show a downward increase in density as they approach the mantle. (b) Section through the continental crust based largly on seismic data. Compared with the oceanic crust, continental rocks are of lower density and higher silica content.

Together these rocks form the Earth's oceanic crust, on average 7 km thick. Intruded into the lower levels are great numbers of igneous dykes, mainly of basaltic composition. The average composition of this crust is also basaltic.

Weight per cent of oxides in the Earth's oceanic and continental crust

Crust	SiO_2	Al_2O_3	Fe_2O_3	FeO	MgO	CaO	Na_2O	K_2O
Oceanic	48.7	16.5	2.3	6.2	6.8	12.3	2.6	0.4
Continental	60.2	15.2	2.5	3.8	3.1	5.5	3.0	2.9

Within the regions of oceanic crust are major topographic features such as oceanic ridges, to which reference has already been made. Together these form an 80 000-km-long chain of submarine mountains rising from the abyssal depths. Current volcanic activity is characteristic of the crestal regions of these, and indeed it is here that new oceanic crust is being generated. The ridges rise between 2 and 3 km above the ocean floors, and often are offset by lines of fractures which may be traced for several thousand kilometres. These are termed *transform faults*. Groups of volcanic islands also are common. Some of these are found in linear groups that are associated with fractures or plate margins. Other chains of islands, like the Hawaiian–Emperor group, are located well away from both ridges and fractures; these are a particularly striking feature of the Pacific Ocean. Many of these are volcanically and seismically active.

Continental crust

The continental crust stands proud of the ocean basins and is also much thicker generally. It is much easier to study than the oceanic crust, for obvious reasons. It averages 35 km in thickness, but may be between 60 km and 80 km; beneath mountain ranges are regions of 'thickened crust'. The oldest sample of continental crust so far discovered is about 4000 million years old. The continents are also covered by large areas of sedimentary rocks whose thickness varies between very thin to more than 15 km.

Isostatic and gravity data both indicate that the upper part of the continental crust is composed of relatively non-dense material, sometimes termed *sial*.

Such sialic material has an average density of between 2.7 and 2.8 g/cm³, and appears to be dominated by rocks rich in silica and aluminium (hence SiAl). The most abundant rock types are sandstones, silicic gneisses and granites. Generally speaking, the gneissose types tend to be found at greater depths. Below these metamorphic rocks are believed to be gabbroic rocks akin to those found beneath the ocean floors.

The continental rocks are buoyed up on the denser materials beneath. The actual boundary between the crust and mantle is defined by a seismic horizon known as the Mohorovicic Discontinuity, or 'Moho' for short. Within this zone earthquake waves experience a sharp increase in velocity from 6.8 to 8.1 km/s, indicating a change in composition of the rocks encountered, those below the Moho being peridotitic in type – much richer in magnesium than even the oceanic crust. Geologists believe that such a rock composition approximately represents a residue of the fractional crystallization processes that generated the oceanic and continental crusts from mantle material.

The Earth's gravity field

If the Earth were uniform in density or a uniformly-layered sphere, its gravity field would also be spherical and decrease radially away from the centre of mass, as required by the inverse square law. Our planet is, however, neither spherical nor uniformly layered, thus its gravity field varies from place to place. The Earth's gravitational constant was first measured in 1798 by Henry Cavendish; he came within 1.2 per cent of the modern figure. He also was the first to calculate the mass: 5.97×10^{24} kg; from this he found the density – 5.52 g/cm³ – considerably greater than the density of most surface rocks.

The reference surface for all gravity measurements is called the *geoid*. This is most easily understood by assuming the Earth's surface to be uniformly covered by water: the surface of the ocean would define the geoid surface.

By observing the way in which orbiting satellites are affected by the mass of the Earth, it can be shown that this imaginary surface shows considerable departure from anything as simple as an ellipsoid. Map makers, who are particuarly concerned about the exact form of the geoid when trying to establish the precise latitude and longitude of points on the Earth's surface, have settled for a calculated solution, known as the 'spheroid'. Broadly speaking, it corresponds to an average geoid.

Points that lie in the same latitude have a theoretical gravitational attraction that is solely dependent upon latitude. By measuring the Earth's gravity field with sensitive instruments called gravimeters, geophysicists can determine the actual gravitational force at any point and, after making a number of corrections, can arrive at a figure that can be compared meaningfully with the theoretical gravitational value for the same point. The disparities between the computed and measured values are called *gravity anomalies*.

When gravity anomalies are considered on a worldwide basis, it is found that they are generally positive over the oceans (meaning there is an excess of gravity over the predicted value), while on the continents, particularly over the sites of high mountains, anomalies are negative (deficiency of gravity compared to that predicted). The higher the mountains, the higher the anomalies become. This confirms what has been said in the previous section about the nature of the oceanic and continental crust, i.e. the rocks beneath mountainous continental regions are considerably less dense than those beneath the ocean floors and beneath lowland regions.

By studying gravity data from thousands of stations spread all over the Earth's surface, gravity maps have been drawn. These allow geologists to define

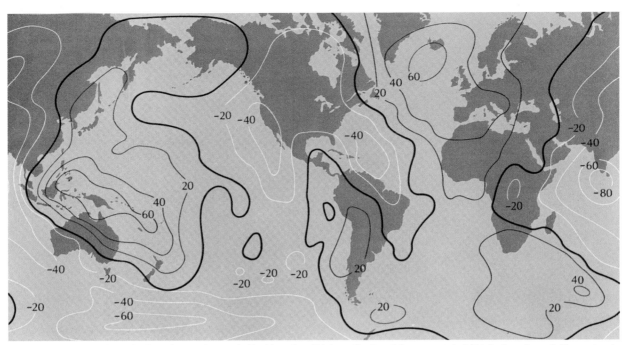

World gravity map. Gravity varies from place to place on the Earth's surface. This is a reflection of the asymmetrical distribution of mass in the crust and mantle layers. The geoid is the term for the shape which most closely approximates the ocean surface. Departures from this ideal form are indicated on the map by regions of positive and negative gravity.

regions where unusual gravity values occur, and stimulate them into trying to explain what this means in terms of the internal workings of our dynamic planet.

The structure of oceanic regions

In recent years a variety of techniques has been developed which has led to a dramatic improvement in our understanding of the ocean floors. Gravity data collected from orbiting satellites, sonic mapping of the sea-floor, and a new kind of aerial mapping, called geotectonic imaging, in which very accurate measurements of the height of the sea surface are made by satellite-borne radar altimeters, have all greatly enhanced our knowledge.

Modern theory states that the lithosphere is divided into a series of rigid lithospheric 'plates' which move slowly over the more ductile asthenosphere beneath, rather as do log rafts over a slowly moving river. The rate of relative movement of these plates is of the order of a few centimetres per year. Plate boundaries may be of three types: divergent, where adjacent plates are moving away from one another; convergent, where they are moving towards one another, and conservative (also termed passive), where adjacent lithsopheric slabs slide past each other.

At divergent boundaries, such as those defined by mid-oceanic ridges, the lithosphere is being penetrated by basaltic magma which has risen from the upper mantle as convective motions within this have pulled the brittler lithosphere apart. Rifts and fractures occur along ridge crests and it is via such weaknesses that basaltic magmas rise, flowing out on to the ridge flanks where eventually they will solidify into new ocean crust. Since the sea-floor is known to be gradually spreading away from such axes, divergent plate margins are also called *spreading axes*.

Because the lithosphere is relatively brittle, a great deal of strain is imposed on it by the convecting mantle below. Convection cells do not have straight boundaries, and therefore to accommodate the strains imposed by the irregular patterns in the mobile mantle, the oceanic crust fractures along *transform faults* which trend roughly perpendicular to oceanic ridge axes. Such a trend is normal to the faults which open along the ridge crests, and it is often found that where the two kinds of fracture intersect, dome-like masses of lava form, almost as if new crust is being formed from numerous, fault-bounded feeders.

New crust for old

Now that the picture of the Earth's crustal structure has been painted, it should be evident that beneath spreading axes the asthenosphere may rise to within a few kilometres of the surface. This brings relatively dense, peridotitic material up from depths of between 50 and 70 km to an unusually high level. Naturally, as the magnesium/iron-rich, silica-poor, material moves upwards, the hydrostatic pressure is lifted, and it expands; consequently it becomes more buoyant. Furthermore, since all this is happening beneath an insulating layer, there will be little loss of heat from the rising peridotite, with the result that it begins to melt. Less than one fifth of the volume of peridotite melts during this upward movement, and this will have a basaltic composition since the temperatures attained will be close to the melting points of silicates such as olivine, pyroxene and calcium-rich plagioclase feldspar – the constituents of basalt. The basaltic rock melt will thus accumulate near the base of the Earth's crust.

Over a lengthy period of time, the basaltic melt will lose heat and begin to crystallize. The first crystals to separate from the melt will be those with the highest freezing temperatures, amongst which olivine probably is the most important. Olivine crystals, being denser than the magma in which they find themselves, will sink under the influence of gravity to the bottom of the magma pocket. Later, or even at much the same time, calcium-rich feldspar will join the olivine. Together they form layers of gabbroic rock that comprises the lower levels of the oceanic crust.

Some of the melt, however, may escape through fractures and cracks and rise towards to the surface, where it will be extruded as basalt lava: basalt because the other constituents of the peridotitic host rock have remained below in the gabbroic layer. En route for the surface, a significant proportion of the melt will freeze in feeder fissures as igneous dykes

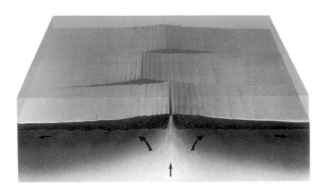

Section through the oceanic lithosphere, showing the rise of basaltic magma derived by partial melting of the upper mantle material into the upper crust. Some magma solidifies in dykes, the remainder reaches the surface, spreading out on either side of the central rift zone. As the lithospheric plates move apart, so the basalts cool, become denser, and settle lower into the oceanic layer.

and sheets, bolstering up the oceanic crust as a result.

As the newly-formed oceanic crust develops, so it is dragged away from its source by sea-floor spreading; in consequence it cools even more, contracts, and becomes denser. Therefore it will sink down below the level of the source region, in an effort to maintain the isostatic balance with the underlying mantle. The depth to which this crust will settle has been shown to vary with the square root of its age. Thus, the older crust, the lower it will have sunk.

Continental structure

Because there is a constant recycling by the Earth of its crustal elements, there is no oceanic crust that is much older than 200 million years. Destruction of oceanic crust occurs at convergent plate boundaries: it is denser than its continental counterpart and is thus more amenable to subduction.

The continents stand proud of the ocean floors; they generally resist recycling and stand imperiously, buoyed up on the denser substrate as slabs of oceanic lithosphere slither beneath them down

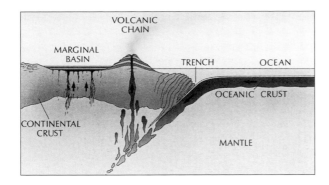

Section through a destructive plate margin, showing the relationship between oceanic and continental crust. Where oceanic crust is being subducted beneath the margin of the continent, some of the downward-moving slab melts, generating magma. Much of this cools deep down, but some rises towards the surface and issues from volcanoes. Very often, behind a volcanic arc a marginal basin opens. This may subsequently be filled with sedimentary rocks and lavas.

inclined planes. This destructive activity is termed *subduction* and the zones along which it operates, subduction zones.

Where a plate bearing continental crust at its leading edge converges on a plate made solely of oceanic material, the denser oceanic slab slides down beneath the leading edge of the approaching continental slab. The sinking oceanic crust is gradually heated up, and will be destroyed by a mixture of heating and mechanical destruction, as it subsides ever deeper inside the hot Earth. In contrast, the less dense continental material stands proud, steadfastly resisting subduction; this is why continental crust yields much older ages than any current sea floor lithosphere. The oldest goes back to at least 4000 million years BP (Before Present).

The rocks exposed on the continents can be subdivided into two main types: (i) accumulations of sedimentary and volcanic rocks of wide aerial extent which show the effects of only minor deformation; and (ii) strongly deformed belts of sedimentary, igneous and metamorphic rocks; the latter comprise what are termed *orogenic* or *mobile belts*. The second group

generally underlie the rocks of the former group, which may not occur everywhere.

The bulk of the continental crust has a rather complex history and has been affected by mountain-building (orogenesis) at some stage or another. Every orogenic belt represents a long history of events, spanning perhaps several hundreds of millions of years. Individually such belts are a manifestation of a variety of geological activities, operating over very long periods. Understanding the story which is locked up in their rocks has been a long-term preoccupation of geologists.

Many features of orogenic belts closely resemble ones of much more recent age; in fact they have similarities with those evolving today. This provides hope that careful study of modern rocks in such environments eventually will lead to a greater understanding of their much more ancient counterparts. One thing we can be very certain about is that orogenic belts which currently are adjacent to one another, have not always been so. Throughout the aeons of time, continents have split apart, moved together, sundered again, collided and undergone a host of different erosional and deformational modifications. The study of continental rocks yields a story of large lateral translations of the Earth's lithosphere; the constant changing in geography is a characteristic of the Earth.

9

The Earth from space

Methods of remote-sensing

It might be argued that the best way to understand the geology of any area is to go there on the ground and see it first-hand. Certainly it has to be said that any geologist worth his or her salt would have a natural preference for doing this, which is exactly why NASA sent the geologist Jack Schmidt on the Apollo-17 lunar mission: they valued the first-hand opinion of a trained scientist amongst the Taurus-Littrow rocks. However, sending people to remote places can be expensive, while sending them to inaccesible, politically-sensitive, or dangerous places could be disastrous, even life-threatening. Then again, when on

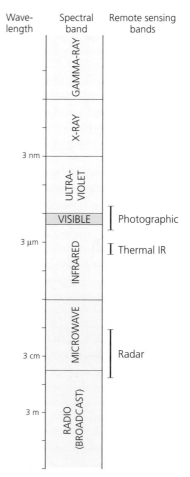

The electromagnetic spectrum, showing remote-sensing bands.

the ground one only gets an impression of the local picture, and it has to be said that sometimes this needs to be augmented by a more synoptic view.

This is where remote-sensing comes into its own. Viewing the Earth from an aircraft or an orbiting satellite does not necessitate entering sensitive or dangerous areas, neither does it restrict observation to a small region; furthermore it can provide synoptic coverage over a long period, and in a range of wavelengths – something that is seldom possible on the ground. Some of the images obtained from spacecraft launched towards other worlds have stunned us with their beauty and surprised us with what our planet looks like from great distances.

The methods utilized to observe the Earth – or any other planet – depend on the way in which matter absorbs, reflects, refracts or radiates electromagnetic radiation. Each can be measured by airborne instruments or scanners mounted on orbiting satellites. Conventional photography can detect only that part of the spectrum between long-wavelength ultraviolet and short-wavelength infrared. Photographs usually are taken on panchromatic film, using special filters to minimize the effects of atmospheric scattering. Colour photographs add a further dimension, and colour infrared film has also been developed which is sensitive to thermal emissions. In the days when remote observations could only be made from aircraft, photography was the principal technique used.

In more recent times, images have been obtained by scanners – like vidicon cameras – and digitized. In this way large numbers of images can be stored in computers, transmitted to Earth and later enhanced or otherwise manipulated to maximize their potential and to suit various purposes. They can then be archived without fear of degradation, something which is a constant worry with photographs.

Multi-spectral scanners have been used more and more in recent times. Such remotely-operated systems employ a mirror to scan a swath of country on the Earth's surface; the collected radiation is then split into discrete wavebands by a series of prisms, passed to one of a number of photocells, and recorded on multi-track magnetic tape. Periodically the information is transmitted back to Earth-based receivers. The variety of wavelengths employed means that scientists can select those images which suit their particular needs, whether it be geology, biology or oceanography.

Another technique employs not solar energy but radiation which is actually generated onboard a spacecraft – something which has been done very successfully by the Space Shuttle. One of the most important of these *active* remote-sensing systems (those which utilize solar radiation are termed *passive* systems) is radar. Such long wavelengths are highly useful, since radar not only provides shadowed images like photography does, but can also penetrate into surface rocks and provide information about the electrical and roughness properties of target materials. The other bonus with radar is that its wavelengths are not hindered by clouds. Radar images can be viewed just as photographs can, and if supplemented by radar-generated altimetric data, can be utilized in producing very high-resolution topographic maps. This technique hinges on the fact that radar, like light, is electromagnetic radiation and, as such, travels at the velocity of light. If a note is made of the time a vertically-directed radar pulse leaves the transmitter, and the time when it is received again (having hit the surface and returned), it is a simple matter to calculate the distance from the satellite to the surface. Radar mapping was of vital importance in the topographic and geological mapping of the cloud-covered planet, Venus, by the US Magellan spacecraft.

Despite all of these newer techniques, aerial photography remains the cheapest and is still widely used. Overflights are normally planned so as to provide overlapping coverage which allows for stereo-

Eleuthera Island, Exuma Sound, Bahamas. This Space Shuttle image shows the low Eleuthera island rising just a few metres above the Atlantic's waters. To the east lie the deep blue waters of the ocean, while to the west are the 10-m-deep shallows of the Bahamas Banks. Here the water is warm, shallow and very salty. These conditions favour the precipitation of the carbonate mineral aragonite (lighter hues) which, in the form of oolites, is being swept across the shallows by tidal currents.

scopic pairs of pictures to be obtained. By using a mirror stereoscope, the interpeter is then able to view regions in three dimensions, a highly informative method much used in planning field surveys. However, as can be imagined, such surveys are only cost effective when small regions are under scrutiny. For large regions, satellite imagery is the order of the day. Multi-spectral, radar and thermal infrared imaging are now commonplace, while measurements of gravity, temperature, pressure and atmospheric composition can all be made by instrumentation mounted on orbiting vehicles. The archive of remotely-sensed information about our world is now immense and has contributed hugely to our understanding of geology, meteorology and biology.

Views of the Earth at visual wavelengths

Satellites such as Landsat and Seasat and orbiting vehicles such as the Space Shuttle have all contributed extensively to our understanding of global geology. Even earlier than these came manned spacecraft such as Gemini, Skylab and Apollo; the images returned by these early probes should not be forgotten. For many people these were the exciting first synoptic views they had seen and, even to trained scientists images of important regions such as the Red Sea and Afar Triangle, generated a huge amount of excitement. Many of these images were taken by hand-held cameras, allowing the astronauts to choose when conditions were good for obtaining photos, and to indulge their own particular predilection with regard to features and areas of interest.

Since missions were quite widely separated in time, it became possible to follow any changes which had affected particular regions. For example, many images were taken of the sand seas of Arabia and northern Africa; and it became possible to trace changes in the type and pattern of dunes which had occurred. It also allowed volcanologists to observe

where new eruptions were occurring, or trace the movement of ash clouds sent up after major eruptive events. One particular instance of this was the eruption of the volcano El Chichon, whose ash cloud was seen spreading out into the stratosphere and moving around the Earth under the influence of the westerly winds. In this chapter a selection of visual images are presented.

Views of the Earth: the weather

Earth's normal weather occurs in the troposphere. This is the lowest atmospheric layer and it extends upwards from the surface to between 11 and 16 km. Winds are generated due to differences in atmospheric pressure from place to place. There is a general upward movement of heated air along the Equator and also at latitude 60°; where such upward air movement occurs, the atmospheric pressure is relatively low. The general wind circulation is very much a function of this pattern and of the Coriolis effect, a function of the Earth's rotation. Thus, in the northern hemisphere, the winds flowing towards the Equator blow from a northeasterly direction, while those at the same latitudes in the southern hemi-

Satellite image of cumulus clouds over the US east coast. [Photo: NASA]

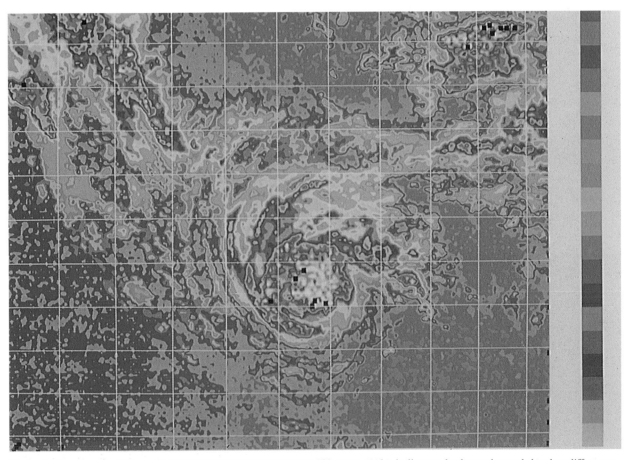

Nimbus-3 false colour composite image of Hurricane Camille, 18 August 1969. This storm, 400 km in diameter, has been colour-coded to show differences in temperature: cool air shown in blue, through yellow, to warm air in red. Wind speeds in the eye of the storm reached 175 km per hour. [Photo: NASA]

sphere blow from the southeast. The same kind of modification of wind pattern affects air movement at higher latitudes, where the principal movement of air is from west to east. Along the equatorial belt itself, there is a zone of very gentle winds, known to mariners as the Doldrums.

On a more localized scale, the movement of air around a centre of low pressure (a cyclone, or depression) is clockwise in the southern hemisphere and anticlockwise in the northern. The reverse is true of high pressure (anticyclonic) systems.

In the southern hemisphere the wind circulation at low levels is more consistent than it is north of the Equator. This is largely because there is so much ocean while, in the north, the continents have a much more concentrated development.

Before the age of space exploration, meteorology was to a large extent an uncertain science. Reliable weather prediction in essence depends upon receiving reports quickly from widely separated points on the planet's surface. Even today neither sea- nor land-based meteorological stations are evenly distributed over the globe. The whole situation was changed

when it became possible to survey the Earth from high altitude, particularly from dedicated weather satellites. Since the 1960s these have proliferated and now play a vital rôle, not only in weather prediction, but also in understanding Earth's weather systems.

Earth in the infrared

The electromagnetic radiation recorded by photographic satellites originates in the Sun. The Earth does, however, emit radiation of its own, but this is of long wavelength and has to be collected by instruments specially built to measure energy in the far-infrared; these are called radiometers. This thermal radiation can be reemitted solar energy, geothermal energy released in the vicinity of volcanoes, or man-made energy, such as that put out by power stations or in areas of urban population. Detection is now a fine art; witness the infrared guidance systems used in modern missiles and the methods used to gain night-time intelligence during recent military campaigns.

Thermal infrared measurements are normally made at night so that the swamping effect of reflected

Quantitative night-time infrared image of the Hawaiian volcanoes, taken in February 1973. This is a specially-processed image which shows the summit of Mauna Loa (left) and part of the Southwest Rift Zone with pit craters along it. The thermal characteristics of the different volcanic rocks are nicely picked out in different colours, thus those with the lowest radiant temperatures are shown in black, the highest in white. Note the relatively high temperatures along the crater rims, due to these being built from relatively dense lavas which contrast with the more porous flows of the caldera floors and flanks. [Photo: NASA]

solar radiation is minimized. Thermal images look rather different from visual ones and are extremely useful for vegetation surveys, for distinguishing differences in rock type and for analysing the thermal output of volcanically-active regions. Most images are computer-generated and colour-coded to bring out differences in emission levels.

The Earth viewed by radar

A great deal of interest was aroused by recent Space Shuttle missions, not least among the reasons being that onboard one of the early *Columbia* flights was an experiment known as SIR-A, an acronym for *Shuttle Imaging Radar-A*. Although this apsect of the mission ended prematurely, the problems subsequently have been ironed out, and highly successful radar imaging has since been achieved.

Pulses of microwave radiation (between 1 mm and 1 m wavelength) transmitted towards the Earth can penetrate clouds and haze, and can therefore reach the surface of the Earth unimpeded. Such radiation will be reemitted from the materials it encounters in different ways, each particular material having its own radar 'signature'. The material properties that affect this signature are the surface roughness, the dielectric constant – a measure of a material's insulating properties – which increases with increasing moisture content, and the slope of the ground. Smooth surfaces appear dark on radar images, while rough surfaces appear bright or speckled. A good radar image will highlight the topography of an area and therefore enhance linear and other structures of interest to the geologist.

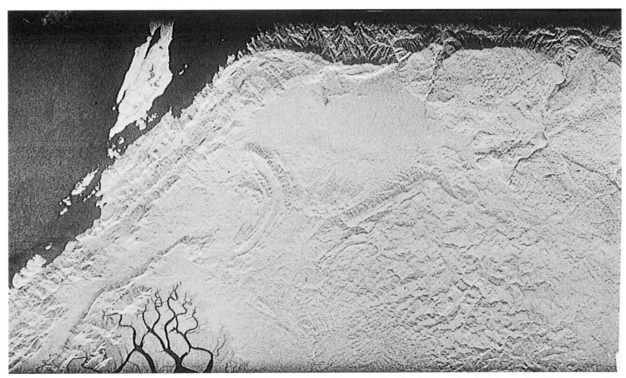

Radar image of the Kalpin Chol and Chong Korum Mountains of Xinjiang, China. This was obtained by the Space Shuttle *Columbia* in 1981. It shows striking linear features which are the result of folding and faulting of the rocks during the Tertiary and Quaternary. [Photo: NASA]

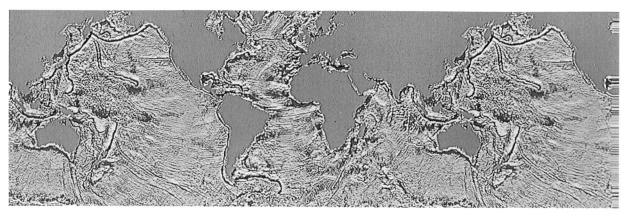

Seasat gravity map of the world's oceans.

Seasat's view of the oceans

In 1978, NASA launched a satellite called Seasat, onboard which was a sensitive radar altimeter capable of making very accurate determinations of the height of the ocean surface. After three months of operation this failed, but not before it had surveyed the oceanic regions between latitudes 72° north and south. It provided scientists with what is called *geotectonic imagery*.

The measurements made, accurate to between 5 and 10 cm, can be processed to reveal striking images of the ocean floors. The basis for such a conversion is the fact that the sea height is directly related to the strength of the Earth's gravity field at sea level, thus the water piles up in regions of high gravity, and sinks down where the field strength is low. The gravity field is at least partly dependent upon sea-floor topography, so the information approximates to a topographic chart of the deep-sea-floor. Such a chart strikingly picks out many linear features, for instance deep trenches, ocean ridges and transform faults.

Part three

Patterns of Earth history

Pounding surf attacking the limestone cliffs at Ulu Watu, Bali, Indonesia. A part of the cycle of change which has constantly affected the Earth: mountains are built, forming land; the sea attacks them and wears them down; they return to the sea; and so on. [Photo: Paul Doherty]

10

Dating the rock record

The Earth has existed for at least 4650 million years, and during that time the heat engine has driven processes which have been responsible for large-scale modifications of the lithosphere and crust. A wide variety of new rocks has been formed; some still exist, others have long since been destroyed and recycled. Those which remain may have been at the Earth's surface for a very long time, others have appeared only recently; of these, some have remained almost unaltered since their emplacement, others have been severely modified.

In trying to understand Earth history, geologists have had to follow numerous lines of enquiry, rather like setting out along the strands of a giant web, weaving their way intricately along a maze of threads in a quest to make sense of an astonishingly complex pattern of events. They have risen to the challenge in many different ways, some using quite simple techniques, others utilizing sophisticated technology; however, whichever method they may have employed, the clues sought will have resided in the rocks themselves, either in their chemistry and physical make-up, or in their mutual relationships with one another, spatially or temporally.

Timing the various events which have shaped our planet is of the utmost importance. This can be achieved by establishing the relative positions of rocks with respect to one another, using the principles of superposition and intersection coupled with palaeontological evidence, or it can be done absolutely, i.e. by radiometrically dating samples, thereby assigning events a precise age. One characteristic of the story which has emerged, is its cyclicity. Similar sequences of events have occurred many times as Earth orbits the mother star. In the following pages I shall attempt to described both the methods used and the story which has emerged.

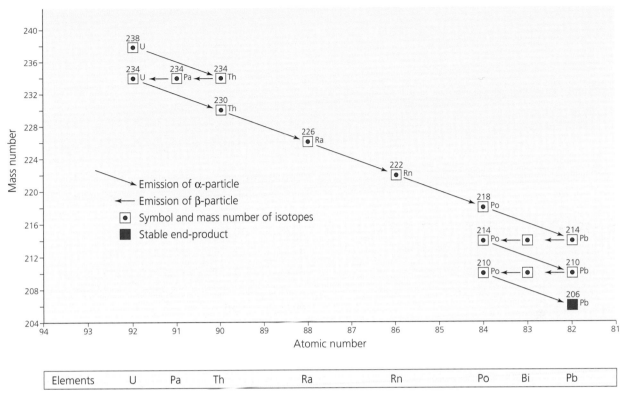

Diagram to show a typical radioactive decay series; beginning with one isotope of uranium (U) and ending with an isotope of lead (Pb).

Radiometric dating

The constituent atoms of each chemical element are held together by electrical forces that pervade the Universe. Each atom consists of a relatively heavy nucleus consisting of one or more positively-charged protons and electrically neutral particles called neutrons. Roughly 99.9 per cent of the atom's mass resides in this nucleus. Circling around it are one or more negatively-charged particles called electrons – which are very light indeed – whose total negative charge is exactly balanced by the positive charge of the nucleus.

The sum of the protons and neutrons in the nucleus is called the element's *mass number*, while the total tally of protons alone comprises the *atomic number*. Atoms having the same atomic number belong to the same chemical element. Those having the same atomic number but a different number of neutrons, are called *isotopes*. All elements have at least two isotopes, some have many more. For instance, the element mercury has seven natural isotopes.

Some elements have naturally-occurring 'parent' isotopes that are radioactive; these may decay sponta-

neously to lighter, 'daughter' isotopes of other chemical elements. Elements which have this property are termed *radiogenic*. An important property of such elements is that any such disintegration always occurs at a fixed rate for a specific isotope, and this fact provides geophysicists with a very powerful tool for dating certain minerals that occur commonly in crustal rocks. It is called radiometric dating and it forms the basis of what is termed *geochronometry*, a method of assigning absolute ages to rock samples.

Uranium has a number of radioactive isotopes; so too does thorium; then there are also argon, lead, potassium, rubidium and neodymium; each has a number of both radiogenic and non-radiogenic isotopes. Taking uranium as an example: the isotope ^{238}U decays by a somewhat complicated process to the non-radiogenic isotope of lead, ^{206}Pb. In so doing the unstable uranium nuclei emit charged helium atoms called alpha particles, as well as electrons – called beta particles – and short-wavelength X-rays. It is these emissions which bring about the changes.

The rate at which decay progresses bears a constant relationship with the quantity of ^{238}U present and is expressed in terms of the isotope's half-life: the time taken for exactly one half of the parent uranium isotope to convert to the daughter isotope. Thus the half-life of ^{238}U is 4500 million years; another of its isotopes ^{235}U, which is also radiogenic, has a shorter half-life, i.e. 713 million years. The isotope ^{238}U makes up about 99.3 per cent of the natural uranium at the present time; the remaining 0.7 per cent being composed of ^{235}U. The much shorter decay period of the latter suggests that the proportion of ^{235}U must have been far higher in the past than it now is. ^{235}U also decays to a non-radiogenic isotope of lead, ^{207}Pb.

Thorium behaves in a similar way; thus ^{232}Th decays to another non-radiogenic isotope of lead, ^{208}Pb. This shows that natural lead includes three radiogenic isotopes, at least; there is also a fourth isotope which is non-radiogenic (^{204}Pb). A significant number of minerals contain uranium, or thorium, or both, and by using a complex instrument called a mass spectrometer, it is possible to find the ratio of radiogenic leads to the parental radiogenic isotopes of uranium and thorium. The ratio provides the basis for calculating the age of the mineral concerned.

Minerals which contain these particular elements are relatively rare; however, other decay sequences occur in argon, potassium and rubidium – all commoner elements than the above. Such elements are found in a wide variety of common rock-forming minerals. Carbon, too, has its uses as a rock dater.

Radiometric dating, as we have seen, is achieved by the study of parent–daughter isotope pairs: rubidium–strontium, potassium–argon, argon–argon, neodymium–samarium, uranium–lead. However, for successful dating, even with modern techniques, the minerals concerned must fulfil certain requirements. Firstly, they must be free of alterations produced by circulating solutions: these would have leached out a proportion of the measurable elements, so skewing the ratio measured. Secondly, metamorphism has the ability to drive out some elements more readily than others. Its effects must be taken into account when such changes have left their indelible mark.

Frequently it is possible to assign absolute age via a number of different decay sequences that have taken place within the sample. There is usually a high degree of consistency in results yielded by the different methods. There are, of course, exceptions; but over the years many of these have been explained away. For example an apparent inconsistency arose when a rock was dated by the ^{40}K/^{40}Ar method: the isotopic ratio derived for a hornblende crystal in the rock yielded an age of 1000 million years. A mica crystal, on the other hand, gave an age of 750 million years. How could this be? The answer lay in the fact that the rock was heated up during a period of metamorphism which took place 250 million years after the rock had crystallized. Because mica has an atomic structure that is permeable to argon, while hornblende does not, all of the radiogenic argon was lost from the mica 750 million years ago, when it was heated up. Subsequently it began accumulating radiogenic argon. By using two individual mineral species present in the same sample, both the age of the rock and the timing of the metamorphic episode could be established.

Isotopes commonly used in radiometric dating

Parent isotope	Daughter isotope	Half-life of parent
uranium-238 (^{238}U)	lead-206	4500 million years
uranium-235 (^{235}U)	lead-207	710 million years
rubidium-87 (^{87}Rb)	strontium-87	4700 million years
samarium-147 (^{147}Sm)	neodymium-143	130,000 million years
potassium-40 (^{40}K)	argon-40	1300 million years
carbon-14 (^{14}C)	nitrogen-14	5730 years

Relative dating techniques

One of the fundamental tenets of geology is that any sequence of strata laid down on the Earth's crust is emplaced with the oldest rocks at the bottom. This is

Sedimentary strata in the walls of the Grand Canyon, as seen from Desert View on the South Rim. The rocks on top of the rim are of Triassic age and become progressively older down to the level of the Inner Canyon, where Pennsylvanian-age rocks sit with considerable unconformity on older strata that extend downward into Pre-Cambrian schists. Note the general horizontality of all of the rocks in the upper canyon walls.

the basis of the Law of Superposition. It stems from the observation that where we see horizontal layers of sand and mud being deposited on beaches or along river floodplains, we never see one layer being deposited beneath a previously deposited bed; every new layer is deposited on top of the previous one. In consequence, a field geologist sampling strata at successively higher levels in a mountain side, should be looking at progressively younger deposits, assuming these are of sedimentary origin. He or she would be able to assign each stratum a relative age simply on its position in the sequence, i.e. older strata would lie at lower levels than younger ones.

However, it has become clear that things are not always quite as they seem. The Earth periodically undergoes phases of intense deformational activity, with the result that sequences of stratified rocks may become displaced from their original positions, folded, faulted and sometimes completely overturned. The upshot of this is that some rock sequences show repetitions of the same sequence, sometimes brought about by faulting, in other instances produced by folding and thrusting. The field geologist thus has to exercise great care while mapping, taking full account of every minute structure and characteristic left in the rock sequence. Experience shows that in some regions the same part of the rock succession is repeated many times, a fact that is readily provable where some distinctive 'marker horizon' picks out the details of deformational activity. Elsewhere, parts of the rock succession are missing completely – due to erosion – quite the reverse of a repetition of the strata!

Some kinds of igneous rocks are stratified in the same way as are sedimentary types. For instance, lava flows tend to be laid down as piles of rock units, often showing similar conformable relationships to those of sediments. Their relative age is therefore quite readily ascertainable. Lavas also may be extruded onto a heavily eroded landscape, infilling pre-existing

valleys or older craters and fissures. Again, the relative ages of the lavas and older rocks are clear to see. Other igneous rocks are intruded into the crust, being seen at the surface only if erosion has stripped off their cover. Intrusive bodies thus often show crosscutting relationships with stratified sequences, or can be seen to have arched them up as magma rises into high crustal levels. It is clear that any igneous rock which cross-cuts sedimentary strata must be younger than the latter, and by the same token, strata arched up by an intrusion must be older than the igneous body which affected them. Relative ages are relatively simple to establish in each of these instances.

Metamorphic rocks – those which have undergone change due either to heating or pressure – can also have their relative ages established. Each phase of metamorphism will produce recrystallization of the pre-existing rocks (whether they be sedimentary, igneous or metamorphic), and distinctive new minerals typically are formed in them. Furthermore, the

(a) Ammonites in a grey shale of Jurassic age. (b) Graptolites, also in grey shale, but of Ordovician age. Both types of animal are of great value in relative dating of strata.

(a)

(b)

deformation accompanying recrystallization normally generates distinctive fabrics in the rocks, most being characterized by a planar arrangement of the new minerals produced. A trained petrologist, after studying hand specimens and thin sections of samples collected in the field, will usually be able to assign relative ages to even the most highly deformed rock sequences and thus establish the timing of events that have affected both an individual sample and an entire area.

The use of fossils in dating

Once life had become established on the Earth (around 3800 million years ago, according to the latest research), there was the potential for organic remains to be buried along with sedimentary matter: a fossil record had begun. Although life did not become abundant until about 600 million years ago or, perhaps more correctly, did not take on a form which was capable of being fossilized and therefore preserved for geologists to study until that time, it had developed beyond the single-celled microscopic stage many hundreds of millions of years prior to that time. Colonial animals built reefs at least 1250 million years ago; their remains are to be seen in *stromatolites*.

Wherever fossils have been preserved in rock strata, palaeontologists and stratigraphers have painstakingly recorded the types and abundance of each species found. Over the years this has led to an appreciation of how life has evolved on the Earth. Some species can be shown to have made gradual changes as they have adapted to changing conditions; others simply have succumbed to the external pressures imposed on them and died out. The more adaptable ones generally survived, the process of natrual selection ensuring a continuation of life on our planet, even when conditions were hostile in the extreme.

Geologists now have a very detailed record of the kinds of fossil that characterized strata of different ages from almost every region of the globe. So good is the record that a trained palaeontologist, on the basis of a single fragment of a single fossil organism, may be able to assign it a relative age in the succession of strata. Where more abundant remains are found, he or she will have little difficulty in so doing. Usually, however, it is not an individual species that enables the age to be pinned down, but the association of different fossil species, called the fossil assemblage. From this collection of remains it may be possible to ascertain such things as the depth of water, the salinity and the climate at the location when the organisms died and settled to the sea floor (or the floor of a lake or river). When the fossil and rock record are taken together and analysed, geologists gradually have been able to establish the palaeogeography of the Earth through the aeons of time.

Late Proterozoic stromatolites at Ellery Creek, MacDonnell Ranges, Northern Territory, Australia. These colonial animals are among the earliest common fossil remains.

Records of a restless Earth

As we have seen, records of the Earth's history are found in the crustal rocks. The elucidation of this story by the study of the strata is called *stratigraphy*, and many conventions attach to it. The basic unit for stratigraphic work is the 'formation', the name applied to a collection of strata which are sufficiently distinctive to form a unit for the purposes of mapping. Each formation will tend to contain its own particular assemblage of fossils and, during the formative years of stratigraphic development, this was taken as sufficient evidence to assume formations extended laterally over vast distances, without perceptible change. The situation is not this simple, as was realized in the late eighteenth century, when the similarity of fossils within the same kinds of rocks was predicted to be very likely simply a reflection of environmental factors. In 1789 the French chemist, Antoine Lavoisier, published a number of diagrams which showed that he evidently appreciated the effects that changes in the relative level of land and sea had upon both sedimentary deposits and the life forms they contained as fossils.

Near-shore or 'littoral' deposits, formed in shallow turbulent water, tend to be coarse-grained and to contain life forms adapted to cope with these difficult conditions. In contrast, off-shore or 'pelagic' sediments, which form in deeper water, are relatively fine-grained and contain rather delicate bottom-dwelling organisms, as well as free-swimming or floating forms. It is easy to appreciate, therefore, that deposits of the same age may look quite different and also may contain dissimilar fossils.

If we are to understand what a series of strata really means in terms of past conditions at the Earth's surface, we need to describe not just their vertical position in the rock sequence, but also their lateral distribution and variation. The lateral changes a formation undergoes are a reflection of changing sedimentary *facies*. The latter may be defined as the characteristics of a sedimentary rock which indicate its particular depositional environment and distin-

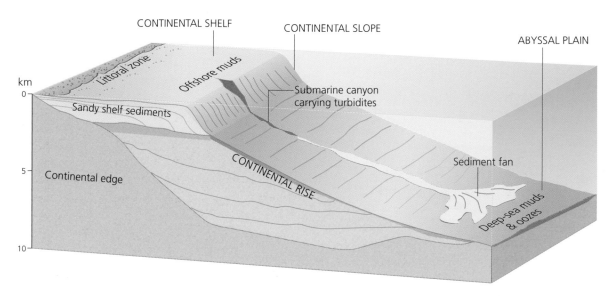

CONTINENTAL SHELF   CONTINENTAL SLOPE

ABYSSAL PLAIN

km
0

5

10

Littoral zone
Offshore muds
Submarine canyon carrying turbidites
Sandy shelf sediments
CONTINENTAL RISE
Sediment fan
Continental edge
Deep-sea muds & oozes

Diagram showing the change in facies encountered in moving off-shore from a continental margin. Coarse-grained littoral deposits pass laterally into finer-grained sediments which have accumulated in deeper water and away from the turbulent waters of the near-shore environment.

guish it from other facies in the same rock unit. An example of facies variation is the passage from pebbles, through sand and silt to mud, encountered in moving progressively off-shore from a modern shoreline.

A bore hole sunk through a thick section of marine sediments will almost certainly reveal a varying sequence of rocks. Most of the variations encountered will reflect the effects upon sedimentation of minor fluctuations in the relative levels of land and sea. Broadly speaking, in marine sequences two kinds of pattern may be recognized: transgressive or 'on-lap' sequences, which reflect a relative rise in the level of the sea compared to the land, and regressive or 'off-lap' sequences, which are the result of a relative lowering of sea level.

Gaps in the record

Major punctuations in the stratigraphic record are caused when a region of the crust previously receiving sediment is uplifted for a lengthy period. This can happen when fold mountains are formed, or when magma rises into the upper levels of the crust, doming it up. The uplifted and deformed crust will then be subject to erosion and gradually older and older strata will be worn away, eventually, perhaps, being

resubmerged as a result. Rock debris from the uplifted region will be transported towards the sea and there will be deposited as sedimentary strata. At a later stage the ocean may encroach on the eroded stumps of the older deposits and lay down fresh sediments which show a marked angular discordance with the older rocks, creating what is termed an *unconformity*. The famous Scottish geologist, Sir James Hutton, predicted that unconformities should occur between older and younger strata. After a painstaking search during the two years following 1787, he found what he had been seeking, a major unconformity in Arran, Scotland. Subsequently he discovered the more famous unconformity at Siccar Point, where older, steeply-dipping rocks were overlain with obvious angular discordance, by flat-lying sediments of lesser age

There are varying degrees of unconformity, depending on what has come to pass during the period represented by the gap in the stratal record. Where there is no angular discordance between older and younger rocks, though a significant time gap is known to occur, the term disconformity is used. Situations like that described by Hutton at Siccar Point, where there is a marked angular discordance, are termed angular unconformities.

Unconformities, being boundaries between rocks,

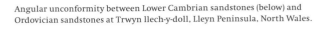

Angular unconformity between Lower Cambrian sandstones (below) and Ordovician sandstones at Trwyn llech-y-doll, Lleyn Peninsula, North Wales.

can be mapped across large areas. During field mapping operations, particular care has to be taken to establish the relative ages of strata both immediately above and below the plane of unconformity. Only in this way can its age be determined. Often the time interval represented by the unconformity changes from place to place; indeed, in some places it may peter out altogether, passing laterally into a conformable sequence.

Impacts and mass extinctions

Now that we have a pretty good idea of the pattern of evolution and the relative order in which different genera and species appeared as time progressed, we are in a position to identify any periods when conditions appear to have become particularly difficult for whole groups of animals or plants. Thus, it is now well established that at the close of Cretaceous times, the previously dominant group, the dinosaurs, very rapidly died out, along with several other groups. Such an event is termed a *mass extinction*. Detailed pal-

aeontological work has revealed that there were similar mass extinctions at or near the end of Cambrian times (about 515 million years ago), at the close of the Ordovician (about 439 million years ago), around 367 million years ago during the latter part of Devonian times, at the end of the Permian (245 million years ago) and at the close of the Cretaceous (65 million years ago). Currently humans are bringing about the most recent extinction period, many plants and animals succumbing to the pressures inflicted upon them by our own species' activities.

Interestingly, most of the mass extinctions had disastrous consequences mainly for tropical marine faunas and floras that inhabited shallow habitats. One explanation for this fact is that a runaway greenhouse effect developed at these times, preventing the biosphere from absorbing excess carbon dioxide. This being so the average temperature may have risen by 2 or 3 °C – quite sufficient to cause devastation among those species which were already living close to the threshold of existence. It is a fact that the processes on which life exists work best just below the temperature at which they would collapse.

Much attention has been given to the mass extinction which occurred between the Cretaceous and Tertiary periods, at what is called the K–T boundary. The fascination of this particular episode rests with the fact that, almost at a stroke (in geological terms) the hitherto most successful group of the Mesozoic era, the dinosaurs, was simply wiped out. From which animals and plants died and which survived (about 90 per cent of Cretaceous species succumbed), it can be deduced that a rise in temperature was the most likely climatic change that occurred. The organisms which became extinct were those which were particularly susceptible to an increase in temperature, because thay were already living in high-temperature environments, or their eggs were laid in such environments, or their large body size was unable to dissipate their body heat efficiently.

The layer of reddish clay at the K–T boundary, South Africa. [Photo: Patrick Moore]

A considerable body of evidence is being amassed which indicates that a major asteroidal impact may have been at least a part of the cause of this devastating event. An early clue to this possibility came when, in 1978, Louis and Walter Alvarez were studying a distinctive clay layer at the K–T boundary. They detected a concentration of the rare element iridium (Ir) at least 30 times greater than in the preceeding and succeeding layers. The significance of this particular metal is that it is very rare in the crust of the Earth, but is found in much greater abundance in chondritic meteorites. Similar iridium-rich layers have been located at other K–T locations on several continents.

Subsequently a search was mounted to try to locate the crater which this projected impact produced. Eventually geologists distinguished two craters: the Manson Crater, in Iowa, measuring 32 km across, and the 180-km-diameter Chicxulub Crater, located in northern Yucatan, Mexico. The fact that two cavities were excavated implies that the bolide

broke up during transit through the terrestrial atmosphere. The large size of both craters indicates that it must have been of very large size, its impact instantly vaporizing the asteroid, sending vast clouds of ejecta, water vapour (from the ocean) and carbon dioxide (from the vaporization of limestones), which were sent up into the stratosphere. There would have been total darkness for several weeks. Eventually the debris would have fallen back to Earth, creating an iridium-rich muddy rainstorm that left its mark in the clay layer the Alvarezes had located.

During the last couple of years, geologists have made the suggestion that earlier impacts may have been responsible for previous mass extinctions. There is certainly evidence in China that such may have been the case. However, this topic is extremely contentious, more conservative scientists being highly sceptical of the whole idea. Despite such doubts, it has to be said that impact events may have had far more to do with the changing course of evolution than once was suspected. Catastrophic theory certainly appears to have had a boost with the work being done on the Chicxulub and Manson Craters. Other factors which might have contributed to extinctions are intense volcanic activity, the splitting apart or colliding of continents (which build mountains and therefore change the climatic patterns) and, of course, mankind.

The cycle of events

The end product of the study of strata and of igneous and metamorphic rocks is the elucidation of geological history over large regions of the Earth. Maps showing the lateral distribution of different rock types, the traces of unconformities and the position of dislocations such as faults, are the basis for description of both local and regional geology. Cross-sections through carefully chosen points will aid appreciation of the vertical as well as lateral variations the rock

THE CYCLE OF EVENTS

units show. When dealing with geology on the continental scale, likely it will be appropriate to draw up special maps showing facies variations, or even the thickness of particular formations that are believed to have major significance.

There are innumerable problems with reconstructing the Earth's history, not the least being that of timing the events which have occurred. An unconformity may not be of the same age everywhere, nor may a thick sequence of sedimentary strata, nor a series of lava flows. Geological processes usually produce extremely slow changes and they do not necessarily act universally. For instance, although we know that the Alpine mountain chain was raised in Europe dur-

ing Cenozoic times, we have no reason to think similar mountains were formed in say, Central Africa, at this time.

Despite all the obvious problems, it is possible to discern a distinct cyclicity in the geological record. Similar progressions of events, taking perhaps hundreds of millions of years to run their course, can be recognized. The geological cycles we now recognize have been discovered by studying not only the sedimentary strata but also the magmatic and metamorphic rocks that occur alongside them; for these yield vital clues to what has taken place on our constantly evolving planet.

11

Integrating Earth history

Introduction

Early in the twentieth century, two European geologists, J. J. Sederholm and Arthur Holmes – the former working in Finland and the latter studying the ancient rocks of Africa – came to the same conclusion regarding the Earth's continental crust, namely that it bore the ingrained record of successive orogenic belts, each having contributed to its growth. The Phanerozoic rocks of Europe record clear evidence for three successive belts, one having developed in northwestern Europe during the Early Palaeozoic, a second in southern and central Europe during Late Palaeozoic times, while a third evolved in the Mediterranean region during Mesozoic and Tertiary times. Subsequent research has shown that by studying the rocks associated with each belt, it is possible to recognize patterns which may have affected regions far removed from Europe, offering a blueprint for the various stages in orogeny and crustal accretion.

Individual orogenies had their own distinctive characteristics, these being related to the events which occurred during a particular period of time. There is, however, a tendency for major orogenic episodes to be grouped together in time; furthermore, within such groupings the events which take place, at least superficially, are of similar type. As an illustration of this, we can cite the Caledonian and Hercynian orogenic cycles, which were related to continental collisions and the accompanying closure of ancient oceans. In contrast, during the Mesozoic the principal cycles were a function of continental rifting and sea-floor-spreading events.

Naturally, each orogenic cycle shows pulses of waxing and waning activity, the development, establishment and eventual degradation of such a belt being an extremely slow, lengthy process. It seems that the typical life span of most belts is of the order of 200 million years or more; it may be four times as long.

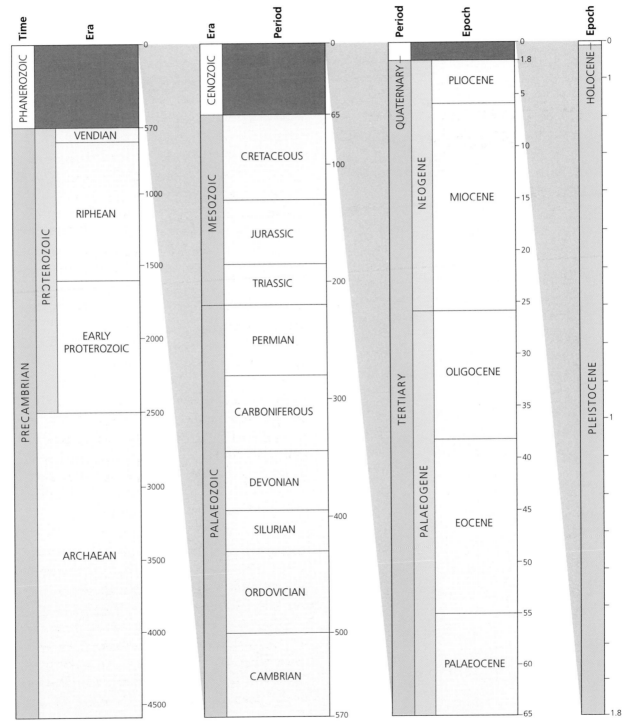

The geological time-scale.

Grouping the geological cycles

The sequence of events which characterizes any of Earth's mobile belts has variously been termed a 'geological' or 'orogenic' cycle. Such cycles have recurred throughout at least the past 2500 million years and have overlapped both in time and space, making their interpretation often difficult and problematic. Typically a cycle includes extensive magmatic activity, both deep down and at the surface; this may peak at certain times, giving clusters of radiometric ages for the igneous rocks produced. After studying the distribution of igneous rock ages for all continents over most of geological time, some geologists believe it is possible to discern peaks in global magmatic activity, notably between 2800 and 2600 million, 1900 and 1600 million, 1150 and 900, and 500 million years ago. This is not to say that activity did not occur at other times, but does suggest that tectonism has been pulsatory.

The early geological cycles of the Archaean and Proterozoic eras.

		m.y.
	EOCAMBRIAN	700
LATE ERA (PROTEROZOIC)	GRENVILLE OROGENY	
	KEWEENAWAN	
	NAZATZAL OROGENY	1300
EARLY ERA (PROTEROZOIC)	PENOKIAN OROGENY	1600
	ANIMIKEAN and HURONIAN	
ARCHAEAN	ALGOMAN OROGENY	2400
	TIMISKAMIAN	
	SANANAGAN OROGENY	3000
	KEEWATIAN	

Geologists have recognized three gross subdivisions of time: (i) the Archaen, which includes all events prior to 2500 million years ago; (ii) the Proterozoic, which includes those cycles occurring since the end of the Archaen and prior to 1000 million years ago (although there is some debate about the precise timing of the latter date); and (iii) the Late Proterozoic – Phanerozoic (up until the present).

Orogeny and magmatism

We have seen that along certain active zones within the ocean basins, new basaltic crust is being generated. Evidence from magnetic striping and radiometric dating indicates that newly formed crust, together with the lithosphere beneath it, will then spread at rates of between 1 and 10 cm per year, away from the spreading centre. In many large ocean basins – although best seen in the Pacific – the oceanic crust is being drawn down into the mantle, eventually to be reabsorbed there at depths of around 700 km.

The return of oceanic lithosphere to the mantle takes place along inclined zones which reach the Earth's surface in deep trenches that scythe through the ocean floors. These represent convergence zones where two lithospheric plates come together. One of the plates – the one whose leading edge is composed of oceanic crust – is being deflected below the other. The inclined zones down which this underthrusting occurs is known as a subduction zone. Friction between the adjacent lithospheric plates gives rise to frequent earthquakes whose depth of focus increases towards the overriding plate. The sloping, seismically active plane is called a Benioff Zone.

As the oceanic crust is carried back into the mantle, the sedimentary cover along its leading edge is mostly scraped off and highly deformed, while the oceanic material starts to be partly remelted at depths ranging between 100 and 300 km. It may, however, continue to subside largely intact to greater

View from the volcanic island of Lipari, showing Salina, Filicudi and Alicudi – four members of a group of seven islands belonging to an active island arc which began to form in the Tyrrhenian Sea 330 000 years ago.

Granite boss standing out from the desert scrub at Arkaroola, South Australia.

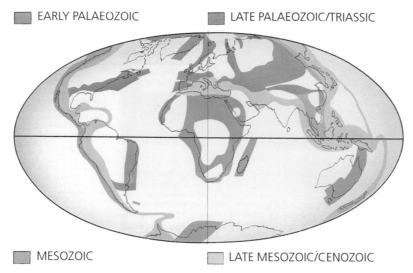

☐ EARLY PALAEOZOIC   ☐ LATE PALAEOZOIC/TRIASSIC

☐ MESOZOIC   ☐ LATE MESOZOIC/CENOZOIC

Map of world showing major orogenic zones.

depths, finally breaking up at around 700 km. Because it is substantially less dense than the mantle material surrounding it, the magma produced as the sinking slab melts rises towards the surface and is erupted as lava which often emerges to form chains of volcanic islands. These are called *island arcs*: Japan and the Aleutians islands are two such examples.

In some cases subduction of oceanic lithosphere takes place directly adjacent to a continent – as currently is happening along the western margin of South America – in which case some of the magma generated rises through the sialic material of the continental margin, emerging as volcanoes built high into cordilleran ranges, e.g. the Andes. The remainder of the magma will consolidate as huge *batholiths* which rise into the mountain root zones, there to cool as vast granitic intrusions.

It is perhaps now more apparent why orogeny and magmatism go hand in glove. Plate tectonic theory helps to explain how plate movements can account for the variety of geological phenomena we see, especially those associated with mobile zones. Seismic activity can also occur without volcanism. A particularly famous example is that of the San Andreas Fault zone, where plate movements are 'passive', i.e. two plates are moving alongside one another. This does not involve generation of magma, but it does mean that periodic major earthquakes occur. Major mountain-building events generally are produced where

plate convergence involves continental crust, as was the case with the Alps, the Himalayas and the western cordilleran chains of the Americas.

Cratons and shields

If we take a very broad look at a geological map of a typical continental mass, such as North America, we at once see that it is built from different components. Extensive regions of the interior, especially in Canada, are quite flat and have remained more-or-less undisturbed since Pre-Cambrian times; at the most these have been mildly warped or subject to up-and-down (*epeirogenic*) movements. Blocks of such ancient crust are termed *shields*. When analysed in detail, it is found that such shields consist of a number of Pre-Cambrian mobile belts that have been welded together and now form extremely resilient crustal units. Indeed, ancient shields form the core regions of all of the Earth's continents. The orogenic belts of the oldest parts of cratons are difficult to interpret, but is quite clear that those of Proterozoic and younger age, are the result of plate collisions involving subduction. The rocks within these once mobile zones are the only clues we have about Earth's ancient, long-closed, oceans.

Most of the rocks of a shield typically are highly metamorphosed, indicating that they have at some time been deeply buried and involved in orogenic

Satellite image of the Pilbara region of Australia. This shows ancient Proterozoic cratonic blocks (light colours) which are separated by narrow zones of strongly-deformed rocks belonging to ancient orogenic belts (greenish colours).

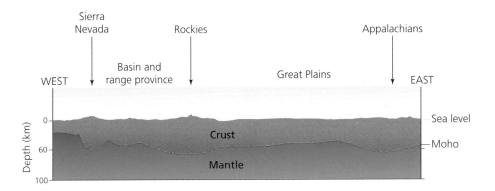

A diagrammatic section through the North American continent, showing the general structure of a typical cratonic region.

movements. That they currently lie at the surface shows that the shield has been eroded over the aeons of time, such that the once deep-seated rocks now outcrop at the surface. The typical rocks of a shield sequence are granites and highly metamorphosed sedimentary rocks which have long since been converted to schists and gneisses. Alongside these are an assortment of much-deformed volcanic rocks. Such an assemblage, often yielding very ancient absolute ages, indicates that there already was extensive orogenesis in Pre-Cambrian times.

Still focussing on North America: in other regions of this continent the shield rocks are covered by very old sedimentary strata which show only minor effects of either deformation or metamorphism. The sediments probably accumulated in subsiding basins formed in response to fracturing or downwarping of the ancient crust. These old strata, which range in age from Pre-Cambrian to Palaeozoic, provide us with vital records of the Earth's early history, including life. The sediment-covered and uncovered shield areas comprise what are called *cratons*. Each of the Earth's continental regions has a cratonic core.

The lifespan of an orogenic belt

A typical orogenic or 'mobile' belt is associated with continental, cratonic rocks. The body of evidence which geologists have amassed indicates that the deformed rocks so typical of these regions originally collected in huge basins which, early in the evolution of a belt, received vast thicknesses of sedimentary

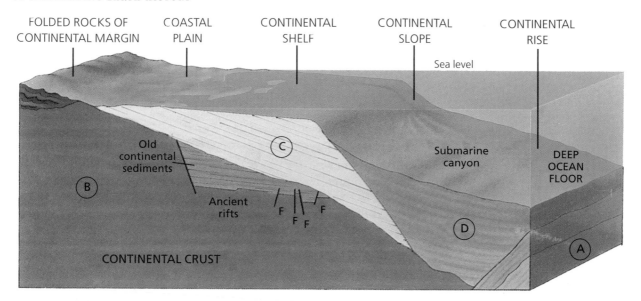

Diagram to show the development of large basins adjacent to a continental margin. A slab of oceanic lithosphere (A) is being subducted beneath a plate bearing continental crust (B). Along the continental margin, land-derived detritus accumulates in a sedimentary basin (C) – initially wedge-shaped in form – which overlies rifts in the continental crust below. This sediment builds up on the continental shelf. In deeper water, further from the margin, marine sediments and volcanic rocks accumulate to produce a thick prism of sediments which underlie the continental slope and deep-ocean floor (D).

debris. Apparently this built up wedge-shaped bodies which, in some cases, reached a total thickness in excess of 5 km! These sediment-laden troughs are sometimes called *geosynclines*, a term which now is not much in favour, but was once a major item in all geological literature. My personal preference is for the term *marginal basin*, since this quite clearly describes the setting of such regions of sediment accumulation, i.e. at the continental margins.

When two lithospheric plates converge, bringing blocks of continental crust together, the crust experiences severe compression, and the accumulated marine sediments of the sea-floor adjacent to the continental margins become caught up in this regime. As if trapped between the jaws of a massive vice, the sedimentary pile is squeezed as the crust is shortened, becoming deformed into piles of folded strata which often are broken up by low-angled faults called *thrusts*. At the same time, new mountains are thrown up along the site of the subducted ocean floor and, during later stages, folded and faulted strata may be thrust over the shield margins. Accompanying such movements is the intrusion of large granitic masses (*plutons*) which solidify to form batholiths.

Once new mountains have been constructed, nature immediately begins to tear them down (equilibrium conditions being a completely flat terrestrial surface, albeit with spherical shape). Erosion produces massive deposits of blocks, boulders and sand which become distributed along the margins of the young, rising, mountain chains. As mass is removed from the mountain massifs in this way, isostatic readjustments gradually occur, the eroded mountains slowly rising to compensate for the loss of mass. Such mountains will continue to rise for a lengthy period, as are the modern Himalayas. In some cases a chain may experience later rejuvenation, the ancient deformed roots being eventually exposed high up amid peaks of newer mountains. The Alps and Appalachians provide us with two examples of this phenomenon.

Exactly how long does all this take? Naturally it takes a very long time, but with the gradual refinement of our radiometric dating procedures, and advances in our knowledge of the stratigraphy and deformation history of the Earth's mobile belts, we can be more specific than that. The Alpine belt of southern Europe appears to have formed over a period of 200 million years, and is still evolving today. The Caledonian mobile belt, whose rocks are found in Europe and Greenland, took much longer to evolve, about 450 million years to be precise. The North American cordilleran belt spans a period of over 750 million years. Evidence suggests that some of the very ancient Pre-Cambrian belts took even longer to evolve into their final form.

The Earth's primaeval crust

What was the Earth's earliest crust like? We certainly cannot see remnants of it anywhere today, but we can use evidence from our neighbouring world, the Moon, and from other terrestrial-type planets. The most ancient Moon-rock has an age of around 4600 million years; so far the oldest terrestrial sample goes back only to 4000 million years. Prior to that time we can only assume that the Earth's crust developed in much the same way as did the Moon's.

After accretion an enormous amount of heat was built up inside the Moon and it is generally assumed that the outer layers, at least, became entirely molten, giving rise to what is termed a *magma ocean*. Within this ocean those crystals which eventually formed first settled out from one another, the dense magnesium/iron-rich mineral, olivine, sinking downwards, while the calcium-rich plagioclase, floated upwards. The latter gave rise to the Moon's first true crust, made from the mineral anorthite (calcium-plagioclase), and called anorthosite. Much later (around 3800 million years ago), the denser, olivine-rich magma that formed the lower levels of the magma ocean, rose into a number of huge impact basins that had been excavated in the anorthositic crust, congealing there to form the lunar maria.

Initially the primitive lunar crust was rather thin and frequently it was pierced by volcanic eruptions. The lavas produced spread out over the primordial crust, adding to its volume, until eventually it thickened and became reasonably stable. As time progressed it thickened so much that it became more difficult for magma to rise through it. However, the thickening process was slowed down by the constant bombardment by impactors, large and small. Impacts continually punctured the surface, spreading out mantle-derived lavas which were then recycled by intense convection in the mantle layer. Today, the intensely cratered ancient crusts of the Moon, Mercury and Mars are the closest replications of early Earth that we have.

Assuming that a similar process occurred on Earth, then we believe that the earliest crust was not anything like the oceanic crust that eventually developed; neither was it like continental material. Rather the primordial crust was probably basalt-like in character – because it was produced directly from Earth's mantle – and only gradually was it modified by the processes of fractional crystallization. It took a long process of refining before the composition of the most primitive crust changed sufficiently to produce recognizable oceanic and continental crustal types.

Among the oldest continental rocks are those of Greenland, where granitic rocks and sediments yield

What the Earth's early crust may have looked like (Apollo image). The margin of Mare Serenitatis (dark and smooth) is made from fluid basaltic lavas, the adjacent highlands (light) are heavily cratered and composed of anorthositic breccias. [Photo: courtesy of NASA]

ages of 3800 million years. The sedimentary rocks contain grains of the very resilient mineral zircon, which evidently was inherited from even older continental-type rocks, since radiometric dating of individual zircon crystals yields an age of 4200 million years. Grains of the same age have now been identified in Western Australia too. The mineral zircon is found only in continental rocks – it is never found in mantle-derived basalts – and its presence can also be noted in 3960-million-year-old rocks from northern Canada. The zircon data clearly imply that continents existed on our planet at least as far back as 4200 million years.

The cratonic rocks of Greenland and Canada might be considered out-of-the-ordinary; however, similar Archaen cratonic remnants are also to be seen in Antarctica, in Australia and in South Africa. There is no doubt that early continents had been formed in a period of around 400 million years. We now need to imagine a process whereby the early 'scum' of the Earth could become modified sufficiently for it to resemble the granitic cratons with which we are so familiar.

Going back to our impression of the primitive crust for a moment: in the early years of Earth's evolution the mantle undoubtedly was hotter than now, and convecting much more strongly. Its lower viscosity would engender more numerous and smaller convection cells than now exist, meaning that the Earth's surface had a thin crust underlain by large numbers of vigorously-convecting cells. In some measure this might give rise to the 'spotty' kind of structure akin to that currently seen on our neighbour world, Venus, where large numbers of cells are believed to underlie corona structures.

Under a typical upwardly-convecting cell the thin primordial crust would fracture, volcanoes building up lavas that were derived from the underlying mantle. Each time this occurred, the new lavas would take on a slightly different composition from the mantle,

due to the process of partial melting and fractional crystallization. The base of the crust would also be plastered with similar lava. In time, following repeated eruptions of this type, the crust would locally thicken until its base became sufficiently depressed that it began to melt. Again, due to the fractional crystallization principle, the less refractory elements would tend to be concentrated in the melts produced. These lights elements would therefore tend to rise into the thickening crust, thickening it further. Should the volcano eventually become so large that it rose above sea level, then erosion would further modify the crust.

The processes operational above rising cells would slowly modify the mantle composition, but not to such an extent that it resembled modern continental materials. The particular rock assemblages of Pre-Cambrian times which we believe represent what was produced occur in what are called *greenstone belts*. These now comprise folded sequences of dark volcanic rocks, shales and impure sandstones called greywackes which are rich in volcanic fragments.

The opposite situation to the above would have been found where the convection cell began to move downward at its edges, or more precisely, where two downwardly-moving adjacent cells met. Here the limbs of the cells would cool, sinking and trying to drag down the primordial crust which had formed. However, the crust would be too light to succumb, therefore it would become thickened and compressed. The upshot is that a downward bulge would be formed again, with accompanying melting at the base of the crust and injection of this lighter material into the overlying crust.

So far the scenario is similar to that of the upward-moving cell, but here the similarity stops. In the case of the downwardly-moving convection cells, no new mantle magma rises into the crust (as it did with the upward-moving cells), so no mantle-derived melt would be present to redress the compositional imbalance. Therefore, as more and more activity followed, there would be a relatively more rapid modification of the crustal composition than occurred above rising cells. As the modified crust thickened it would became more buoyant and would eventually rise above sea level, wherupon erosion would further move the composition along the road to a continental-type chemistry. If the granitic crust actually developed in this way, then it becomes relatively simple to explain why the oldest cratons are configured such that granitic cores are surrounded by greenstone belts.

The oldest dated rocks

As one might expect, the sensible places to seek very old rocks are the cores of the continents. Until very recently the oldest dated samples came from Isua in Greenland (3824 million years) and the Slave Lakes region of Canada (3960 million years); however, eclogites from the Roberts Victor Mine in South Africa register an age of 4000 million years, as do similar samples from Tanzania. More recently zircon ages of 4200 million years have come from Australia. The rocks of western Greenland come from the Isua region, where gneisses exposed after the relatively recent retreat of the ice cap are associated with a variety of highly altered sedimentary rocks, chemically-

precipitated ironstones and cherts, together with volcanic rocks which clearly were laid down under the sea.

Attention has also been focussed on greenstone belts, first brought to light after fieldwork in the gold-bearing Witwatersrand region of South Africa. The African greenstones are ancient volcanic and marine sedimentary rocks which have been engulfed in granite magma. An age of 3500 million years has been established for the greenstone lavas, which have been turned green by recrystallization induced by regional metamorphic activity. Within the belt several cycles of eruption and sedimentation can be recognized. The greenstones are overlain by a series of sandstones and conglomerates that appear to have been generated by the erosion of sialic crust; the greenstones, however, show no evidence for such an origin. The combination of sialic and greenstones sequences has been discussed in the preceeding section.

The greenstone lavas, which have a very low silica content (indicating a very primitive composition?) are interbedded with cherts, and often show pillow structure typical of lava which erupts under the sea. Clearly oceans existed on the Earth this long ago. A rather distinctive textural feature of these lavas is spinifex texture, a distinctive fabric indicative of rapid quenching from very hot magma (1600 °C). Lavas of this type, called komatiites, are significantly different from modern ocean-floor basalts and have also been predicted as primitive lavas on Mars.

Lastly, recent dating of diamond-bearing rocks from South Africa establishes that the diamonds, while they were ultimately derived from the mantle, must have been incorporated in continental-type crust at least 3500 million years ago. From what we know of the depths at which diamonds form, this implies that even as long ago as this, at least 120 km thickness of lithosphere had evolved and become

The Earth's earliest rock? A 4000-million-year-old eclogite from the Roberts Victor Mine, South Africa.

chemically distinct from the underlying mantle. Since that time the continental crust has concentrated some 25–50 per cent of the Earth's heat-generating, radioactive elements, e.g. uranium, thorium and potassium.

The first signs of life

In the earliest aeons of Earth's history conditions were quite unsuitable for life. When the primary atmosphere was stripped away by the Solar Wind, high radiation levels would have destroyed any life that had developed. Later, when volcanism degassed the secondary atmosphere, initially it would have been rich in carbon dioxide, together with water vapour, hydrogen sulphide, carbon monoxide and hydrogen. There would have been very little free oxygen, so that a modern animal, even a lowly one, could not have survived. Things may have been different under the sea, however, and it is not impossible that primitive marine organisms such as algae or bacteria may have appeared quite early in time.

Until recently the oldest life-forms for which positive evidence was forthcoming were presented by American palaeobiologist, J. William Schopf, who showed that bacteria-like microfossils, much like modern pond scum, existed on the Earth 3460 million years ago. However, in November 1996, a team of scientists from the USA, Australia and England presented isotopic evidence for life going back even further, to 3850 million years ago. By studying tiny carbon inclusions in a rock collected from Akilia Island in southern West Greenland, with a high-resolution ion microprobe, the team found a ratio of 100:1 for the isotopes ^{12}C to ^{13}C. This may sound rather uninteresting, but the light isotope of carbon is 3 per cent more abundant than would be expected if life-forms were not present in the sample. This is described as a 'signature of life'. The carbon was found in the phosphate mineral, apatite, often

formed by micro-organic activity – although it is also formed inorganically. The life-form discovered was probably a simple micro-organism, although its shape cannot be ascertained due to subsequent deformation and metamorphism.

Although the above data do not unequivocally prove that life existed at that time, the evidence points in that direction. To obtain definite proof of early life we have to turn to stromatolites, the fossilized remains of primitive blue-green algae preserved in calcium carbonate. Algae, of course, unquestionably are plants; seaweeds are modern examples. Stromatolites date back around 3500 million years, having been collected from a place in western Australia so remote that it is called 'North Pole'!

Even more positive evidence comes from the same region, but from somewhat younger rocks formed 2800 million years ago. In these are found fine, filamentous structures which usually are called pro-algae, and which almost certainly are biological. In the Lake Superior region of Canada we find slightly more complex life-forms, some of them not unlike the blue-green algae of modern times. Conditions on Earth evidently had changed by this time and, although there could not have been much free oxygen in the atmosphere, there must have been sufficient to generate a layer of ozone which shielded the surface from dangerous short-wave radiations from space.

Then, in rocks 1400 million years old, comes the earliest evidence for organisms with more advanced cell structure, called eukaryotic. Much larger cells then appeared quite quickly and, by around 600 million years ago, the oxygen level had reached about 7 per cent of today's value. This was the start of what is termed the Phanerozoic period when abundant, albeit primitive, life developed in such a way as to withstand the rigours of geological processes and become fossilized.

12

The early continents

Early cratons and Pangaea – supercontinent

Pre-Cambrian history is less easy to interpret than younger periods; for a start, the rocks are often deformed and then there are few if any fossils to help us. Even when life began it left no detectable remains. Nevertheless, it seems that the ancient rocks of the continental shields date back at least to 3800 million years ago. Continents existed very early on (as did oceans, of course). From our studies of these ancient rocks, we believe that the sialic layer had a thickness of between 25 and 40 km, compared with the present thickness of between 10 and 70 km. Continents probably were initially quite small, although by around 3000 million years ago, ancient North America possibly was at least half as large as the modern continent. The same may have been true of the others.

The presence of actual continents (as opposed to continental crust) is proved by the detection of quartzites, the undeniable products of the weathering and erosion of sialic crust which has risen above sea level. Such rocks are found, for instance, at Isua in Greenland, having been dated to 3800 million years BP (BP = Before Present). They have been detected on all of the other continents too, but yield different ages from place to place.

When we come to consider the origin and evolution of continental crust, we find conflicting opinions; however, one scenario was presented in Chapter 11, suggesting that convecting mantle plumes generated greenstone belt assemblages over rising currents, while cratonic materials developed above downward-moving cells. The magmatic refining processes which achieved this situation must have been rather slow, yet accounted for major cratonic slabs by no later than 600 million years after the Earth was born.

Apparently the continents grew slowly, new crust being added at the continental margins; in other words, the continents grew by accretion. Rocks from

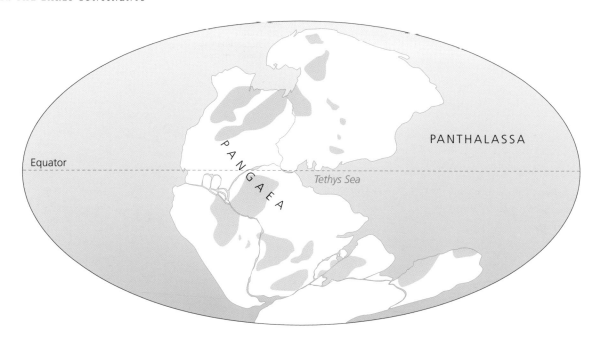

One possible reconstruction of the supercontinent of Pangaea. The ancient continental cores or 'shields' are shown in yellow.

all the ancient shield areas have now been radiometrically dated, the age range being between 2500 and 4200 million years. Generally speaking the oldest rocks are rather restricted, and it may be that the first major continent-forming episode commenced around 2800 million years ago, since we find a cluster of absolute ages of this date amongst the continental rocks of the Earth. However, preserved from this early stage are the remnants of about a dozen small Archaean cratons ranging in age between 3000 and 4200 million years and forming the cores to the modern continents.

The rate of continental growth can be traced by studying the changing composition of the sediments which were won from them and deposited onto the sea-floor. Had all of the continents appeared early on, one would anticipate little change in the composition of sediments since that time. In reality we find that during the Archaean, sediments were won mainly from oceanic lavas (or a more primitive type of crust), a condition which prevailed until around 2000 million years ago. Furthermore true continent-derived sediments first appear at different times on different continents.

We know a great deal more about the more recent rocks than we do about Archaean and Proterozoic ones; however, with the advent of radiometric dating methods and palaeomagnetic measurements, it has become possible to track down how the continents moved about. There seem to have been several quite small continents during Archaean times – presumably kept this way by the hot, rapidly-convecting mantle and the thinness of the primaeval crust. But what about during the Proterozoic: were there several large continents, or just one huge one?

We cannot be sure; there may at times have been only one, a supercontinent whose late Proterozoic successor has been named Pangaea. Another alternative is that there were two: Laurasia (in the northern hemisphere), composed of the cores of the present northern continents, and a southern supercontinent called Gondwanaland, comprised of the present southern continents. Stratigraphic methods alone cannot tell us the complete story; but when we draw on palaeomagnetic data, isotopic dating and structural studies, we can begin to form a fairly confident idea of what happened during these early stages of Earth's evolution.

Palaeomagnetic data suggest that there were in the late Proterozoic (1000–600 million years BP) five independent continents, corresponding to what we now call North America, Europe, Siberia, China and a continent which no longer exists as an entity: Gondwanaland. Proterozoic 'North America'

The principal orogenic events to affect the North American continent during Pre-Cambrian times.

included much of the present continent plus north-western Britain, west Norway, Spitzbergen and possibly parts of northeastern Siberia. Earlier (say, 1100 million years BP) what was later to become North America and Gondwanaland may well have been part of a single continent, while even further back in time, say 1500 million years BP, these two and Eurasia may have been combined, bringing us back to the notion of a true supercontinent.

If this Proterozoic supercontinent of Pangaea did exist, and most of the evidence points this way, it did not persist. There were periods when Gondwanaland and North America split apart, only to re-unite later. Then, many hundreds of millions of years later, around 350 million years BP, Pangaea reformed once more! Mountain building, rifting and the opening and closing of ancient oceans, have all played a part in our planet's history; this is a story I shall now endeavour to tell.

Ancient North America

Ancient North America was the direct predecessor of the modern continent. Currently it has several components: a triangular-shaped block of crust called the North American craton, and two much younger belts of fold mountains – the Appalachians and Western Cordilleras, together with the fold belts of eastern Greenland and northern Canada. The northern part of the ancient craton comprises the Canadian Shield, where we find Archaean rocks; similar rocks extend to the south but are blanketed by younger strata. A feature of the cratonic core is its stability: it has remained a stable region for at least 2000 million years.

The Canadian Shield has been carefully studied, even though parts of it are decidedly awkward to reach. Pioneer work was carried out during the mid-nineteenth century, in the region of Lake Superior, Minnesota and Ontario. The main problem with this terrain is its structural complexity: most of the rocks are highly deformed and also metamorphosed, which makes them difficult to interpret. The oldest part of the Shield lies in western Ontario but extends into Montana and Wyoming, now in the USA. In Archaean times, however, Greenland was joined to this landmass, and it is here that we find the oldest rocks, those at Isua. The ancient continent also incorporated fragments of northwestern Britain.

Two main assemblages of rocks characterize the shield: granitic rocks, including gneisses, and volcanic and immature sedimentary rocks formed in greenstone belts. During Pre-Cambrian times the shield was deformed during at least four episodes of major crustal upheaval. One occurred about 3000 million years ago, a second around 2400 million years ago, a third at 1600 million years, and a fourth between 1000 and 700 million years BP. Each of these episodes created orogenic belts, caused rifting, and in

EAST GREENLAND MOBILE BELT

0.25-0.35 b.y.   0.3-0.4 b.y.

1.8 b.y.

1.2-1.5 b.y.   3.6 b.y.

2.4-2.6 b.y.

1.6 b.y.

CORDILLERAN MOBILE BELT

CANADIAN SHIELD

0.8-1.2 b.y.

1.6-2.0 b.y.

2.3-2.7 b.y.

0.3 b.y. - present

1.6-1.9 b.y.

1.2-1.5 b.y.

0.8-1.1 b.y.

LIMIT OF CRATON

APPALACIAN MOBILE BELT

0.2-0.4 b.y.

Map of the ancient craton of North America, showing the various 'age provinces' as established by radiometric dating. In the north the shield is exposed, but further south it is covered in younger strata. Mobile zones border the craton.

(a)   (b)

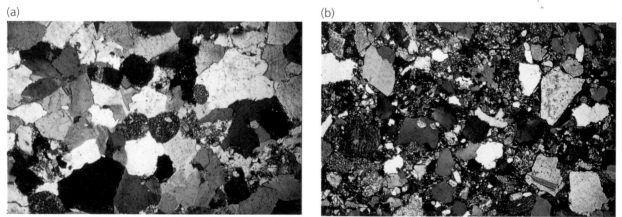

(a) A mature quartz-rich sandstone containing predominantly quartz and rounded grains of siltstone. The rock has become welded together by the out-growth of the quartzes after burial of the rock. Stiperstone, Shropshire; cross-polarized light. (b) An immature, poorly-sorted muddy sandstone, or greywacke. Compared with (a), this sediment shows a greater range in grain size and an admixture of quartz, feldspar, volcanic rock fragments and muddy particles. Harlech, North Wales; cross-polarized light.

The Mazatzal Mountains of Arizona, a range of Pre-Cambrian sedimentary strata, faulted and folded in a highly complex manner. These formed a part of the ancient shield of North America.

some cases collision of lithospheric plates. After successive upheavals the rocks caught up in the mobile belts became stabilized and welded onto the perimeter of the existing continent. Large-scale rising of granitic batholiths into the subcrust helped to stabilize the process. This was so because granitic rocks are of low density and therefore tend to 'float' on the denser lithosphere beneath. The greater the volume of low-density granitic material added to a continent, the more difficult it becomes for it to be subducted. In consequence it remains relatively undisturbed for very long periods.

The sedimentary rocks found within greenstone belts are termed immature, being predominantly what are called *greywackes*. Such rocks include significant amounts of volcanic fragments, together with material won by erosion from the sialic crust; they formed rapidly and in a relatively unstable environment, presumably in the oceans bordering the growing craton. However, not all of the sedimentary rocks originated in this manner. There are other ancient strata which clearly were derived directly from sialic crust, particularly away from the Canadian Shield region. These mature sediments are composed of resilient minerals such as quartz, together with feldspars – both derived from granitic-type crust. The manner in which the grains occur implies that they were laid down in shallow, agitated water, presumably near-shore. It seems that vast shallow seas invaded the early continent, covering extensive regions.

Very late in Pre-Cambrian time, thick layers of marine sedimentary rocks accumulated along the continental perimeter. These rocks now are found in the Appalachians, the Western Cordilleras, eastern Greenland, Norway and northwest Scotland. One suggestion is that these rocks collected in down-faulted troughs (graben) which were formed when ancient North America broke away from ancestral Europe.

The cratonic rocks of Eurasia

Pre-Cambrian rocks are exposed in Scandinavia, the Baltic region, the Ukraine, Britain and in parts of southern Europe; there are further outcrops in Arctic Russia and Siberia. Remnants of ancient crust are found also in eastern China, Manchuria and Korea. Various pieces of fragmentary evidence suggest that the Pre-Cambrian rocks extending from eastern Newfoundland to Massachusetts also were a part of ancient Europe during the Archaen era.

Radiometric dating tells us that the gneiss outcrops of the Ukraine and northeastern parts of the Baltic Shield are between 3000 and 3100 million years old. Overlying these rocks, with marked unconformity, are layers of intensely deformed, metamorphosed volcanic and sedimentary rocks. These in turn were invaded by granites about 1600 million years ago. Apparently the story in this region was one of the alternating splitting and reforming of ancestral Europe, Siberia and China. The European continent also came into collision with ancestral North America from time to time.

The Baltic Shield

This region is undoubtedly one of the most closely studied regions on Earth, at least from a geological standpoint. It can be divided into three parts: (i) a large region of Scandinavia and the Ukraine which is occupied by Early Proterozoic mobile belts; (ii) regions of older gneisses which lie in the north and east; and (iii) a western belt where an ancient mobile

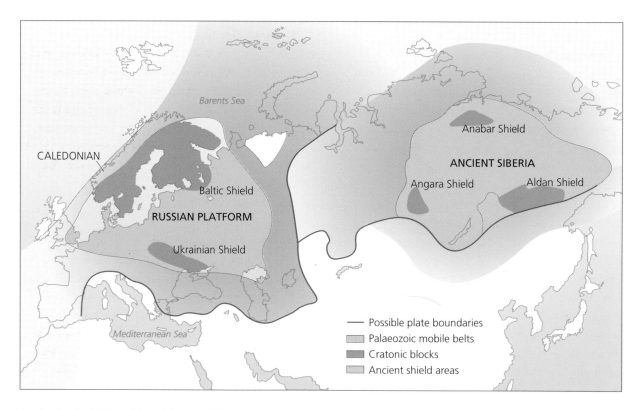

Map showing the relative positions of the main shield areas in ancient Eurasia. The Russian platform and Ancient Siberia were cratonic regions covered by younger sedimentary rocks. The positions of Palaeozoic mobile belts are shown in relation to the Pre-Cambrian cratonic cores.

Map of the Baltic Shield illustrating the various structural elements. The Sveccofennide mobile belts were stable by Early Proterozoic times. These lie to the east of younger belts and to the west of older ones, including the Caledonide mobile belt, which is of Palaeozoic age.

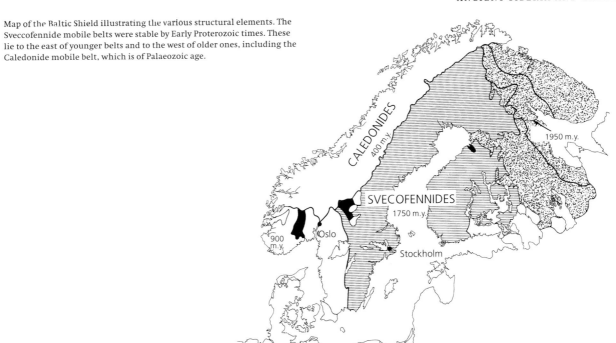

belt has been incoporated into a much younger orogenic belt known as the Caledonides. The first of these regions has extensive sedimentary and igneous rocks dated to between 1700 and 2000 million years BP. They are younger than gneisses which outcrop further north; these yield ages of 2500 million years BP.

The sedimentary strata exposed in Scandinavia and the eastern Ukraine are over 500 m thick, and include quartzite, marble, schist and distinctive iron-rich rocks that frequently are found in these ancient terrains.West of them are found metamorphosed shales and volcanic rocks that are presumed to have accumulated in an unstable marine environments, i.e. a greenstone belt. Around 1700 million years ago the more westerly parts of the shield were caught up in folding and metamorphism, and were invaded by granites. This upheaval may have been the result of collision between ancestral North America and Europe.

Between 1700 and 1000 million years ago, the crust in the region of what is now Finland and Sweden was invaded by swarms of basaltic dykes. Dyke formation is normally associated with extensional tectonics, the stretched crust providing easy access for rising magma from below. The stretching may have been the start of separation between ancestral North America and Europe, but later in the Proterozoic a

chain of new mountains arose at the western edge of the Baltic Shield, indicating not tension but compression. Evidently there had been a change in the plate motions occurring, there being a strong likelihood that a collision occurred between ancient Eruope and either Gondwanaland or North America.

Later in Proterozoic times thick volcanic and sedimentary deposits accumulated in troughs which developed along the continental perimeters; these were deformed during Palaeozoic times. During the late Proterozoic the ocean separating Laurasia from Gondwanaland extended from Novia Scotia, through Newfoundland and central Britain, to Norway, Spain and southern Germany.

Ancient Siberia and China

There are few exposures of ancient rocks in modern Siberia, but there are sufficient isolated gneiss outcrops to conclude that there were a number of shields, the most northerly of which dates back to 3600 milllion years BP. These shields probably formed the core of an early continent. which was later added to by the crushing of orogenic belts against its margins, presumably by continental collisions.

Large tracts of highly metamorphosed rocks occur along the southeastern and southwestern margins of the Siberian craton, having been formed in mobile zones which existed between 1600 and 800 million years BP. Such Proterozoic rocks provide us with clear evidence that events in Eurasia were not dissimilar to those affecting North America at this time.

Small shields are found also in Manchuria, in eastern China and Korea. Although our knowledge of these is not great, it seems that there may have been a number of ancestral Asian continents in Archaean times. The few dates which have been obtained yield ages of around 2500 million years for the most ancient cratonic block.

Gondwanaland

This southern supercontinent incorporated ancient Africa, Antarctica, Australia, Peninsular India and South America. Using all of the methods of investiga-

tion available to them, geologists agree that the pattern of events affecting it was similar to that which affected Laurasia (Eurasia and North America). Periods of relative stability alternated with phases of mobility, during which narrow zones of the crust suffered substantial upheavals. Not only did Gondwanaland show movement with respect to Laurasia, but it seems that the ancient shields from which it was composed periodically broke up and then reassembled in response to mantle motions. The earlier periods of major unrest appear to have peaked at around 3600, 3000 and 2700 million years ago. The oldest samples from Gondwanaland are zircon crystals found in a quartz sandstone from Mt Narreyer, in Western Australia; these have yielded an age of 4200 million years. A little younger are the eclogites discovered in South Africa and in Tanzania; these yield an age of 4000 million years. There follows here a

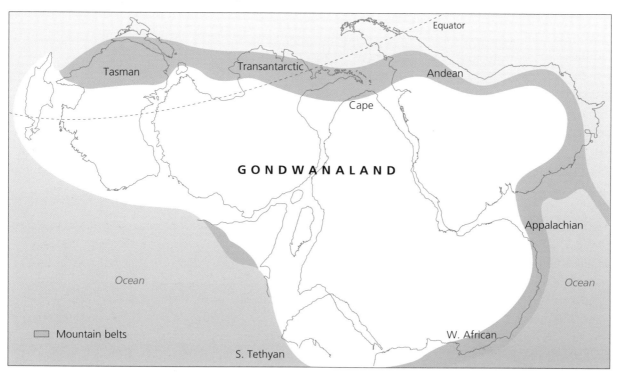

Reconstruction of the supercontinent of Gondwanaland, showing the ancient shield areas which now form the continental cores of Africa, South America, Australia, Peninsular India and Antarctica.

description of the changing fortunes of each member continent.

Africa is home to very large regions of ancient stable crust, although this does not mean that stability prevailed everywhere, all of the time. Indeed, in parts of South Africa, as well as in what now are the continent of Australia and the subcontinent of India, there is abundant evidence for tectonism between 2700 and 2500 million years BP. The remains of ancient fold mountains thrown up at this time are now exposed on the eroded surfaces of the modern continents.

In southern Africa, quite thick sequences of volcanic and marine sedimentary rocks were laid down during the early Proterozoic, which was also a time of extensive rifting. Basaltic magma rose into thousands of fractures, forming dykes which transect the ancient crust; similar dykes formed also in Ghana, where they trend in an east–west direction and have an age of 2400 million years. It could be that this magmatic activity was connected with the splitting apart of Gondwanaland and Laurasia at this time. Certainly there is evidence to support this contention, the most likely line of contact being along the Anti-Atlas Mountains of north Africa.

At the other end of the continent, in the Transvaal, the rock succession includes strata thought to have been laid down under glacial conditions about 2300 million years ago. Palaeomagnetic data show that when they were formed, this part of the continent lay within 30° of the South Pole, from which it follows that much of Gondwanaland must have been undergoing glaciation at this time. Indeed, similar sequences are found in South America, India and Australia.

Following this glaciation, volcanic rocks were deposited above the glacial beds around 2000 million years ago. This accompanied great upheavals which may have been related to subduction of one of these early continents beneath another, or perhaps to collisions between smaller 'microcontinents'. We are

unsure of the details, but it does seem clear that by around 1800 million years ago, North America and Gondwanaland had collided, generating a major new chain of fold mountains.

Five hundred million years later, in the late Proterozoic, they split apart once more. Intense rifting can be identified with this process, not only in Africa, but also in regions as far apart as North America, Europe and Siberia. Later still, thick sequences of boulder deposits and limestones, believed to be glacial in origin, accumulated in parts of South Africa, Antarctica, Australia and northeastern South America. Palaeomagnetic evidence indicates that at this time central Africa and northeast South America sat along the Equator, while almost the whole of the African shield lay within 20° of it. Although Africa was close to the Equator 600 million years ago, parts of it remained glaciated; ice sheets clearly were not confined to the poles at this time in Earth's history.

Ancient Africa

Africa has had a complex geological history. Currently is consists of a huge platform of ancient crystalline rocks which are partially covered by younger strata. Much of the platform is constructed from Pre-Cambrian gneisses, but there are Palaeozoic rocks too. Younger fold mountains occur in the north, on the site of the Atlas Mountains, while there is also a Palaeozoic fold belt in Cape Province, at the opposite extremity. It appears, however, that even while these mountains were being thrown up, the ancient craton remained stable.

The principal regions of Archaean rocks outcrop in the Transvaal, Zimbabwe, Tanzania, Angola and west Africa. Together these build three large shields which have a west–east structural grain. In places the rocks are enriched in valuable minerals such as gold. In many ways they are comparable to the ancient shields

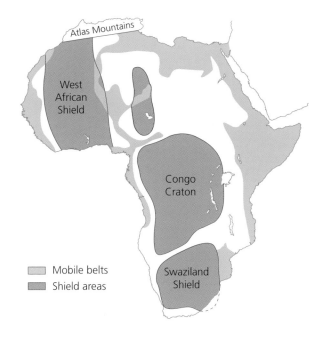

Ancient Africa, showing the Pre-Cambrian shield areas. Each of the principal shields has a structural grain that strikes west–east, and is composed largely of granitic igneous and metamorphic rocks.

West African Shield

Atlas Mountains

Congo Craton

Swaziland Shield

Mobile belts
Shield areas

of the North American Superior Province, but the Transvaal rocks, in particular, are of great interest since they include ancient sedimentary and volcanic rocks that have remained essentially unaltered for over 3400 million years.

In each ancient block the relationships between sedimentary, volcanic and plutonic rocks are complex. As elsewhere, gneiss terrains and greenstone belts occur, and there is a general feeling among geologists that the same geological processes operated in Africa as in Canada and Eurasia. Broadly, it seems that during Archaean times there was a 'basement' of gneissose rocks upon which and against which younger volcanics and sediments accumulated, particularly during orogenic episodes. These arose where the mantle could split apart the existing granitic crust along relatively weak lines. One of the most prominent of these zones separated the ancient gneisses of the Transvaal from those of Zimbabwe; this is called the Limpopo Belt. It appears to have been the site of long-lived crustal activity at least until 2000 million years ago.

Gondwanaland appears to have fragmented about 1700 million years ago. The evidence for this comes from the presence of deep-water sedimentary and volcanic rocks along the Anti-Atlas Mountains of North

Africa; these can only have formed when this part of the supercontinent started to fragment, a significant ocean separating the fragments. However, about 1200 million years ago, Gondwanaland appears to have reunited, only to split again around 1150 million years ago. The last event is suggested by the presence of deep-water sedimentary rocks of this age in the Atlas Mountains. Evidently Gondwanaland and Laurasia once again were rifting apart.

Peninsular India

India is best considered as two regions: the triangular cratonic block of Peninsular India and Sri Lanka, and the Himalayas. The former has remained a stable area for at least 500 milllion years; the latter has been subject to great upheavals throughout Phanerozoic time. The Indian craton collided with the Eusasian landmass in relatively recent times, whereupon the Himalayas were thrown up.

The ancient craton contains rocks which range in age from 2700 to 500 million years old. In the north, old cratonic rocks disappear beneath vast sheets of debris which have been shed by the rising Himalayan mountain chain. Much of India is composed of granitic gneisses which are at least 2000 million years old, but these rocks tend to be highly metamosphosed and difficult to interpret unequivocally. The largest chunk of ancient crust is located in southwest Delhi province; there are other sizeable pieces between Bombay and Delhi.

Several narrow orogenic belts cross the craton, ranging in age from 2000 to 1200 million years. These run along the east coast, southwest from Delhi, and one belt extends inland from Calcutta. The latter is particularly important since it contains substantial deposits of iron ore. Also of interest is the presence of unusual rocks called charnockites, which show evidence of having been formed deep down in the lithosphere.

Peninsular India showing the blocks of ancient crystalline rocks which lie to the south of the Himalayan belt. A large region in the west of the peninsula is covered by younger basaltic lavas of the Deccan Traps.

In the northern part of the peninsula is a vast region of Pre-Cambrian volcanic and sedimentary rocks whose precise age is somewhat equivocal. Structures produced in these strata by ancient currents, show that quartz-rich sandstones and shales were laid down in shallow water, along with some limestone seams. The current structures also suggest that they were derived from a landmass then located towards the southeast. It is possible that they originated on a stable block which existed soon after the old craton became stabilized, at the close of Proterozoic times.

Antarctica

Although 99 per cent covered in ice, geological mapping has been undertaken in Antarctica, largely by means of geophysics. In the west of the continent is a relatively young mobile belt, but to the east, the ice is underlain by a 35-km-thick cratonic block. This East Antarctic Craton is built from gneisses and charnock-ites, many of which have been dated radiometrically. The oldest dates cluster around 2000 million years.

In the coastal region of east Antarctica many rocks yield radiometric ages centring on the period 650–400 million years BP. Here we appear to be looking at older rocks which have been reworked. Rocks of similar age have been found near the western limit of the craton, in the Transantarctic Mountains; some of

these rocks apparently were derived from sediments laid down in a marine trough of Late Pre-Cambrian–Early Palaeozoic age. Thus the history of this region goes back further than the radiometric ages suggest.

South America

A brief glance at a topographic map of this continent reveals that it can be divided into three principal regions: the Western Cordilleras (the Andes), a broad region of plains that lies east of the Andes and narrows eastwards into the Amazon basin, and the eastern high plateaux, separated into Guyanan and Brazilian components by the great River Amazon. Not surprisingly there is a close connection between this tripartite division and the underlying geology.

The oldest rocks are found in the Guyanan and Brazilian Shields; however, there are also smaller ancient blocks to the south, in Argentina. Within each of the main shields is a core of Archaean rocks, mainly gneisses, but also sedimentary and volcanic rocks. In Guyana the ancient gneiss basement is covered by younger pink and red sandstones, shales and volcanics in which contentious 'life-forms' have been located. There are also swarms of sills and dykes which have been dated radiometrically. A minimum age of between 1800 and 2000 million years is indicated for the Archaean rocks here. Other dated horizons yield greater ages (2500 million years BP).

Nearer to the Atlantic coast the rocks have been deformed by a later, Palaeozoic, orogeny, during which substantial plutonic masses of granite were emplaced. After this upheaval, the ancient craton apparently became domed up, forming a broad region of basins into which younger sediments were poured. These sedimentary accumulations have lain undisturbed ever since.

The Brazilian region is underlain by a cratonic block, the ancient gneissose basement extending as

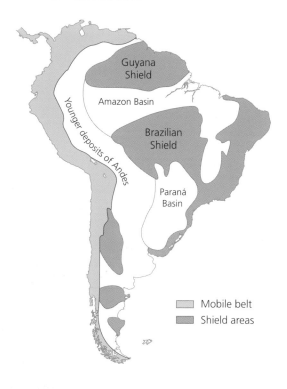

South America. The principal shield areas lie to the east of the Andean ranges. The largest of the ancient crustal blocks are those of the Guyanan and Brazilian Shields whose rocks date back to at least 2500 million years BP. The huge Amazon Basin separates the two shield regions.

The rocks which cover large areas of the craton are of particular interest since they provide a record of almost continuous sedimentation over a 1500-million-year period. As with other cratonic regions, granite–gneiss–greenstone assemblages are typical. Some of the volcanics and sedimentary sequences are, amazingly, almost entirely undeformed, despite their immense age. They appear to have accumulated in cratonic basins that have remained largely unaffected while orogenic movements deformed similar rocks in adjacent regions.

In Western Australia, rocks of greenstone belts have yielded radiometric ages of around 2700 million years (Yilgarn block) and 3000 million years (Pilbara block). Numerous volcanic–sedimentary cycles are represented in these thick sequences, showing that the early history of the continent was not a quiet one. The oldest continental cratonic rocks in the world have been found here: a zircon-derived age of 4200 million years from grains in quartz sandstone found at Mt Narreyer indicates that true continental material existed at this early date.

Igneous activity also affected northern Australia during early Proterozoic times. Radiometric ages of

far south as the River Plate. Both Archaean and Proterozoic rocks are exposed, the oldest being found in the north and east; these yield ages of 2300–2500 million years. Along the coastal belt is a zone of deformed sedimentary and volcanic rocks but these become less disturbed inland. On top of the craton are several regions of Late Pre-Cambrian clastic sedimentary rocks, some of which have a glacial origin.

Australasia

The Pre-Cambrian rocks of Australia are widespread and occur in a large cratonic region which underlies the western and central part of the continent. Younger rocks cover much of the craton, but exposures of the older rocks are good in many areas. This is particularly true of the Yilgarn and Pilbara regions of Western Australia, which are cratonic blocks separated from one another by the Hammersley Belt, an east–west zone of Pre-Cambrian orogenic rocks.

Australia. Vast areas are built from Pre-Cambrian rocks, many of which are of Archaean age. Some of the rocks are of great interest in that they have remained virtually undisturbed since they were laid down, at least 2500 million years ago. They provide geologists with a record of cratonic deposition that is 1500 million years in length.

Pre-Cambrian metasedimentary strata forming narrow ridges in the Gammon Ranges, South Australia. A part of ancient Australia.

between 2000 and 1700 milllion years are typical for plutonic rocks from a wide range of localities. A number of younger igneous episodes occurred in mobile zones developed between the more stable blocks between 1700 and 1000 million years BP, suggestive of the fact that the continent may have been disrupted by repeated movements of smaller microplates within the main cratonic mass.

The Hammersley belt, with a west–east trend, saw accumulation of one of the most remarkable successions to be found on continental crust. Resting on the Archaean basement is a 13-km-thick sequence of volcanic and sedimentary rocks which evidently accumulated in a rapidly subsiding marine trough. Within the sequence is a series of banded, iron-rich rocks which, unlike the rocks above and below them, quietly precipitated on an undisturbed sea-floor. The iron formation extends over an area of at least 140 000 km², thinning away from the central region, suggesting that it was laid down in an enclosed basin that remained undisturbed for a considerable time.

One of the most economically important regions of Pre-Cambrian rocks is that centred on Mount Isa, Queensland. During Late Proterozoic times there was a period of extensive mineralization. This produced the deposits of copper, silver, lead, zinc and uranium for which this region is justly famous. Late in the Proterozoic, glaciation affected the region as it did elsewhere in Gondwanaland, and in several locations boulder-bearing deposits are seen to rest on a striated, glacier-worn surface. Several episodes of glaciation seem to have occurred, peaking at 875, 740 and 610 million years BP. Evidently Gondwnaland traversed the South Pole during this period.

13

The drifting continents

Introduction

A quick glance at a world map shows that if the various continents were cut out like the pieces of a giant jigsaw puzzle, many of them could be roughly fitted together. In particular, the bulge of South America fits quite snugly into the hollow of west Africa. As long ago as the year 1620, the famous scientist Francis Bacon commented that this could hardly be mere coincidence. Not long afterwards, in 1688, R. P. F. Placet, in France, wrote a book in which he claimed that the Old and New Worlds once had been joined together, but had separated prior to the biblical Flood. In the seventeenth century the great age of the Earth was not known, and it was widely believed to be no older than few thousand years.

The Flood was still in the thoughts of the German explorer Alexander von Humboldt, who in 1800 proposed that the Atlantic was no less than a huge river which had broadened as a result of the torrent of water over which Noah had sailed in the ark. Much later, in the 1870s, came a more rational theory by Sir George Darwin (son of Charles Darwin) which involved the Moon. According to Darwin, the Moon once had been a part of the Earth, and had been thrown off as a result of the rapid rotation of the combined mass. This idea was extended, and it was suggested that the hollow now occupied by the Pacific Ocean marked the site of the Moon's departure.

Plausible though this seemed at the time, the idea is completely untenable, if only because the Pacific depression is very slight compared with the bulk of the Moon. It was not until the early years of the twentieth century that the idea of drifting continents was revived. There were several preliminary suggestions, but the real founder of modern theory was a German, Alfred Wegener. His first book on the subject was published in 1912. Although primarily a meteorologist, and also an explorer – he died in 1930 during an expedition to Greenland – Wegener was struck by the sim-

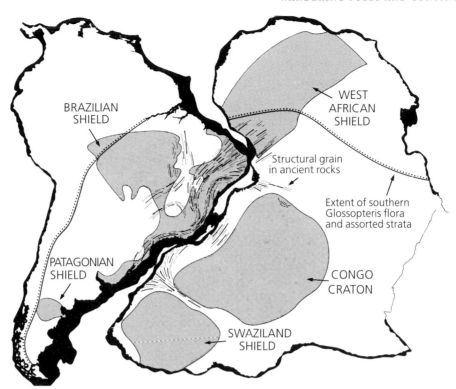

The jigsaw 'fit' between western Africa and eastern South America. This is best achieved along the margins of the continental shelves. Note how geological features on one continent have counterparts on the other.

ilarity between opposing sides of the Atlantic. By using a set of geological matches across the modern oceans, Wegener proposed that a single supercontinent had shattered and drifted apart.

European geologists listened sympathetically but few believed him; on the opposite side of the Atlantic, nothing less than scorn was poured on his head. The idea of a relatively good match between the southern continents did gain some credence, but as to the rest – there really was no concrete evidence. Certainly there were dicrepancies in his interpretation; however, he did point out that there were gross similarities between fossil remains found in ancient rocks in eastern South America and west Africa – just where the two might be fitted together in the world jigsaw. There were other geological correlations, too many to be put down to sheer chance. This left only two possibilities: that there could have been a landmass between the two continents – but there were very many reasons for rejecting such an idea; or that Africa and South America were once joined together.

For several decades Wegener's ideas were almost forgotten in Europe, then, in the late 1950s and 1960s there came a rapid and decisive change in outlook. The Australian geologist, Warren Carey, plotted the continents on a globe and made transparent outlines of them, so that they could be slid around on the globe's surface. Many fitted neatly together. In 1965, Sir Edward Bullard employed a computer to do a similar task; the results were much the same. Finally, as modern techniques were developed and further stratigraphic studies were completed, conclusive support arrived for the notion that the continents had indeed drifted around. Palaeomagnetic data, isotopic dating and stratigraphic palaeontology provided the key data which finally vindicated Wegener. By 1970, even most sceptics were won over; the theory of continental drift had finally entered the realm of fact.

Wandering poles and continental drift

One of the reasons why geologists initially were so sceptical about Wegener's ideas was that there was no obvious reason why the continents should shift around, nor was there any plausible mechanism to achieve movement. An answer to the problem was

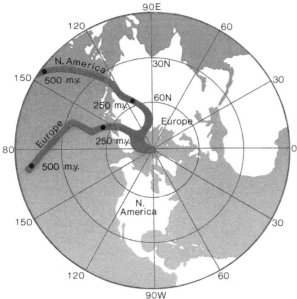

almost perfect fit which can be achieved for some continents which once were joined; consider, for example, the structure of the Saharan Shield, which is around 2000 million years old. The structural grain of the rocks runs north–south towards the interior but then swings west–east towards the Atlantic margin. There is a well-defined boundary between these ancient rocks and younger ones which run into the ocean off the coast of Ghana. If drift is a fact, and Africa and South America once connected, there

first suggested by the great British geologist, Sir Arthur Holmes, who proposed that because the solid mantle was at very high temperature and under substantial pressure, it could actually flow over long periods. Slow motions of this kind would be more than ample to drag along rafts of the less dense lithosphere; although slow, the 'currents' would be extremely powerful. The situation is actually more complex than this, but Holmes certainly was on the right track.

Modern work has established that the lithosphere of the Earth is divided into a number of semi-rigid plates, seven of which are large, and a further five of which are of reasonable size. The boundaries between them are marked by zones of active seismicity and volcanicity. To be precise the term, 'continental drift' really refers to drift of the plates, rather than the continents alone; however, the result is much the same. Drift occurs because the plates are moving relative to one another, and the continents are carried along as part of lithospheric adjustments driven by mantle motions. Once this basic principle became accepted, it was relatively quickly that supporting evidence came along to dot the *i*s and cross the *t*s, so to speak.

The most simple line of evidence comes from the

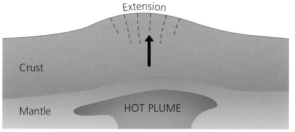

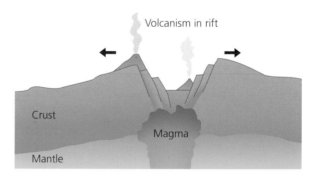

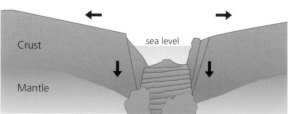

Diagram illustrating the doming of the crust prior to rifting and continental separation. Initially the youthful rift sits high above sea level. Later, as the crust thins and subsides, the rift sinks below sea level, while the dip of the adjacent strata changes to the opposite direction.

should be similar rocks with similar trends on the complementary side of the latter continent. Indeed, this is so; the boundary and a similar structural grain are found in the Brazilian Shield.

Further support for the theory comes from fossil remains. Fossils retrieved from ancient strata in Africa and Greenland, for instance, show that during the Silurian, Greenland was in tropical latitudes, while Africa was in the grip of glaciation! Then again, comparison of index fossils from the Phanerozoic rocks of both Gondwanaland and Laurasia shows that while the two were at one stage widely separated by the Tethys Ocean, at other times they were very close together, if not actually joined. The latter certainly was true between 350 and 220 million years ago.

The most convincing evidence, however, comes from palaeomagnetism, and it was this that finally clinched matters once and for all (although not immediately winning over all geologists). As I have previously mentioned, we can define the positions of the continents with respect to latitude by locating their past positions as shown by their palaeomagnetic imprint. During the late 1950s some curious facts had begun to emerge: as more and more palaeomagnetic meaurements were made from different continents, it was found that, for any individual continent, if the magnetic pole positions were traced for different periods in time, data would not cluster around a single point but would trace out a path across which the pole appears to have passed. This could only be interepeted in one of two ways: either the magnetic pole had moved, or the continents had; initially it seems easier to accept that the magnetic pole position had changed. These paths were called *polar wandering curves*.

Once similar paths had been measured from different continents, it immediately became clear that this interpretation could not be the correct one. For instance, when polar paths for North America and Europe were plotted and compared on a map, while the pole position converges at the present time, 500 million years ago, in the Early Palaeozoic, the poles were far apart. The same was found to be true for other continents too. There could be only one interpretation: the continents had moved. When the ancient pole positions for the different continents were 'put together', they were found to match those positions suggested by Wegener.

The theory of plate tectonics now has come to stay, and it is important for other than purely palaeostratigraphic reasons. Now that we know much better how the outer part of the Earth behaves, it has become easier to track down likely sites for natural resources, such as oil and gas, precious metals etc. Perhaps one day it may become possible to predict earthquakes and volcanic eruptions better. The verification of this theory has been as important to geology as Darwin's theory of natural selection has been to life sciences.

It appears that rifting of continents preceeds drift. In this connection, the geology of the East African Rift Valley is helpful, as well as that of the Red Sea–Gulf of Aden region. In both areas the splitting of the lithosphere was preceded by doming above the rising hot mantle. The brittle crust, having formed an upward bulge, stretched, then fractured, some of the faulted blocks slipping downard to relieve the stress. Because of this, the crustal layer effecively was thinned, with the result that pressure was released on the underlying mantle materials which rose as magmas into the fractured, thinned crust above. Volcanoes were constructed along the weakened line.

During the initial stages, the young rift sat high above sea level; however, if stretching continued for a long period, the crustal layer became so thin that subsidence took place, the crust sinking below sea level. A narrow, shallow ocean was formed, as it has in the case of the Red Sea. Beneath it – and similar young rifted continents – new mantle-derived lavas are rising to extend the area of the oceanic crust on either side of the line of rifting, i.e. the new divergent plate boundary.

It appears that doming often sets up a pattern of triple rifts, one arm of which eventually ceases to be active; this one points away from the newly-formed ocean. When Pangaea broke up it is likely that a whole series of triple rifts were formed within the continental crust, the pairs of active rifts then linking together to form lines of major break-up.

Continental jigsaws

The task of reconstructing exactly how the continental blocks once fitted together at the beginning of the Palaeozoic era is a very trying one. Various attempts have been made to do this and, naturally enough,

there are considerable differences between the results. There are, however, a number of points about which there is general agreeement: North America and Greenland should be put back alongside Western Europe; Africa should be placed alongside South America; the former jigsaw gives us the supercontinent we call Laurasia. Today this is separated from the remnants of the more southerly supercontinent, Gondwanaland, by the broad Alpine–Himalayan chain.

South and west of the Alpine chain are the continents of South America and Africa, while to the east and south are the ancient stable blocks of Australia, Peninsular India and Antarctica. Most geologists

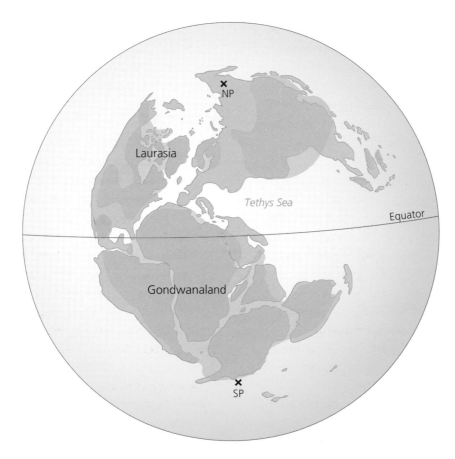

Continental jigsaw. One way in which the various pieces of the continental jigsaw may have fitted together. This reconstruction, drawn for Triassic time, sees Laurasia and Gondwanaland having become joined together towards the west, but more widely separated in the east by the ancient Tethys Ocean.

would restore all these pieces of the jigsaw into one supercontient – Gondwanaland – which existed during the Late Pre-Cambrian, went through a complex history of disruption and reconstruction, and certainly was together again around 200 million years ago. Today widely separated pieces of the Gondwanaland jigsaw are found in the Mediterranean, the Middle East, the Himalayas, Central America and in various parts of the Pacific and Antarctic Oceans.

So, by early Palaeozoic times Gondwanaland evidently existed as a unit; however, Laurasia's pieces do not seem to have then come together. The evidence for this comes from palaeomagnetic data, which shows that during the Palaeozoic the position of Siberia was very different from that of Europe; it was not until much later, in Carboniferous and Permian times, that they became welded together along a mobile belt called the Uralides.

The even larger supercontinent, Pangaea, seems to have existed at least in Late Palaeozoic times, around 200 million years BP; and was coming closer together 100 million years earlier. At the later time Gondwanaland and Laurasia were connected in places, certainly along what is now northern Africa, the north coast of South America and the eastern coast of North America. To the east there seems to have been a much wider separation between the two by Late Palaeozoic times, the great embayment being known to geologists as the Tethys Ocean. This ocean was to remain a feature of the geography for many millions of years. Pangaea, perhaps not with all its components, may have existed earlier than this – something which is still somewhat contentious, but nevertheless a strong possibility. Bearing in mind this distribution of land and sea, clearly it was more a case of universal ocean in which were placed the ancestral continent(s) than the present picture of nicely distributed continents amongst several well-defined oceans.

The fact that this pattern was established meant that the world climate must have been very different then. The great north–south spread of Pangaea would not have allowed the strong west–east air flow which is now established, nor the pattern of ocean currents. To cite but one aspect of climate, the widespread occurrence of desertic 'red beds' and evaporites in Permian sequences suggests that much of the interior of Pangaea was a pretty harsh, arid place. The changing distribution of land, sea and mountain chains had a profound effect on global climate and, of course, the distribution of life-forms

Ancient oceans

The Earth has a distinctly bimodal distribution of elevation. The continents – which occupy about 40 per cent of the total area of the surface (if their continental shelves are taken into account) – lie at an average elevation that is only just above sea level. The oceans, on the other hand, which occupy the remaining 60 per cent of the surface, have an average elevation that is 4 km lower then the continents. The theory of isostacy explains why this is so: the relative buoyany of the crustal materials is a function of density times thickness. The oceanic crust is thin and relatively dense (density = 3.1 g/cm^3); the continental crust is less dense (2.7–2.8 g/cm^3) but thicker. Both types of material 'float' on the denser mantle layer; however, the continental crust stands well proud of sea level as a result of its much greater buoyancy.

Today's oceans appear permanent to us but are relatively recent features. We know that the relatively dense oceanic crust which underlies the ocean basins is nowhere older than 200 million years – youthful on the geological time scale. Plate tectonics explains how the oceanic crust is continuously recycled by the Earth's heat engine, being generated at mid-oceanic ridges and returned to the hot interior via subduction zones at convergent plate margins. Can we learn

Tilted rafts of ophiolite in Oman, Arabian Peninsula. These ancient sediments and sea-floor basalts have been thrust onto the margin of ancient Arabia.

anything about the ancient oceans which have long since been destroyed by plate movements, whose crust has forever been lost and recycled inside our planet?

The first question to consider is what kinds of rocks might we be looking for in our quest? Well, the ocean is floored by basalts erupted under submarine conditions, for which reason many show a development of what is called 'pillow structure' (formed where pillow-shaped blobs of incandescent magma burst through the chilled skin of a lava flow and roll down the sloping sea-floor). We also know that the lower part of the oceanic crust is composed of coarse gabbroic rocks and bolstered by swarms of basaltic dykes. Overlying the igneous crust are marine sediments which, on a modern ocean floor, are thin in the vicinity of oceanic ridges, but thicker away from them. The sediments include shales, siliceous oozes and fine-grained volcanic muds. Towards passive oceanic margins (i.e. those where oceanic and continental crust lies side-by-side and not separated by a plate boundary), thick prisms of marine sediment

Section through a typical ophiolite sequence, showing the relationship between the various rock types. Compare this with the sequence illustrated in Chapter 8.

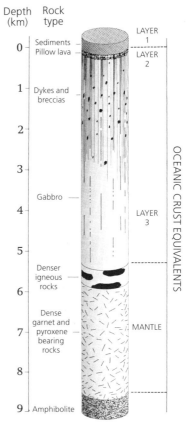

may accumulate. These have a continental provenance. In contrast, along convergent plate margins there are deep trenches in the ocean floor which receive great thicknesses of immature sediment that is a mixture of volcanic material, stuff scraped off the continental margin and huge slumped blocks called mélange. The volcanic material derives from chains of volcanic islands which accompany trenches. These island arc volcanics are not, however, basaltic; their volcanoes erupt andesitic lavas which are mid-way between oceanic and continental crust in chemistry. In the case of both trenches and oceanic ridges, some sediment piles may become unstable, and are swept down ridge or trench sides by fast-moving turbidity currents, generating rocks called turbidites.

Having established what kinds of rocks are formed in oceanic environments, the next question is where are oceans currently being destroyed? The answer is relatively simple: at convergent plate margins, along subduction zones. In such situations, where a slab of oceanic lithosphere is plunging beneath a continental margin, substantial amounts of sea-floor sediments will be scraped off the descending slab and thrust against or onto the continental margin, there becoming highly deformed and often recrystallized. The degree of change wrought might make them unrecognizable in ancient successions. Then again, where a region of oceanic crust is completely obliterated between two colliding slabs of continental crust (as has happened along the line separating Peninsular India from ancient Asia), slices of the remaining sea floor crust may be thrust onto the land by a process known as obduction (the opposite of subduction).

While finding such rocks might appear to be well nigh impossible, in recent times geologists have begun to track down these ancient rocks and, in so doing, to trace out the lines of sutures which mark ancient continent–ocean boundaries. In particular, a suite of rocks known collectively as *ophiolites* has been located among deformed strata within fold chains such as the Ural, Zagros and Appalachian Mountains. Confirmation that these zones are collisional comes from palaeomagnetic data and from observations expanded in the following section. A typical ophiolite rock suite contains altered deep-sea sediments – particularly shales – limestones, silica-rich rocks called cherts, and rocks whose fabric and structure show they were distributed by turbidity currents. Along with these are found submarine basalts, gabbros and some peridotites (although in a very altered state). The current view is that such ophiolites represent slices of oceanic crust, together with some upper mantle rocks, which have been thrust up onto the land during ancient ocean-closing episodes.

Lithospheric plates and orogenesis

Many of the details of plate theory remain to be debated; however, the principle now seems an established part of modern geological thinking. One of the phenomena which has to be explained is the formation of fold mountain chains during periods of orogenesis. Plate tectonics states that orogeny occurs where lithospheric plates converge; this in turn must involve subduction, volcanicity and tectonism. As we have seen, subduction leads to a scraping-off of marginal sediments and the rise of magma into the crust, forming volcanoes. The latter stages of the process see the rocks trapped along the collision zone being compressed and eventually raised up to form new mountain chains.

A deposit often associated with collisional zones of the type described above is mélange. In effect this is a huge submarine landslide, instigated by the tectonic instabilities experienced on the leading edges of colliding plates. Huge slabs of sedimentary and volcanic rocks may literally be scraped off the descending slab, to slither and slide down the trench slope as a complex of coarse and fine debris. Sometimes entire rock

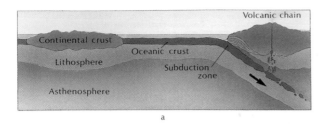

a

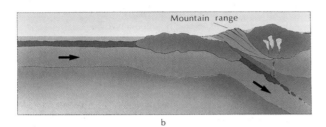

b

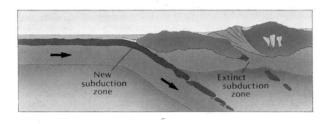

c

Collision between two lithospheric plates. (a) Two plates, one bearing oceanic crust and the other with continental crust along its leading edge, converge. Magmatic activity and deformation of trench sediments takes place. Mélange also forms. (b) Collision between continental crust at plate margins; a new mountain range is formed out of the pre-existing ocean-floor sediments, while andesitic magma rises into the adjacent crust. (c) In some situations the advancing plate may become disrupted, plate motion may cease, and eventually a new subduction zone may be fomed elsewhere.

downward into hotter regions, along with the subducting slab of cold oceanic lithosphere. Large-scale mélange can be found amid the Lower Palaeozoic rocks of western Newfoundland, and also in parts of North Wales.

Mélange belts are accompanied by zones of intense magmatic activity; these usually run parallel to the mélange and arise largely from friction generated during subduction. This causes melting of the upper part of the downwardly-moving slab, as well as of the water-laden oceanic sediment plastered onto its surface. Magmas generated in this way form at depths of between 100 and 200 km and, because they are relatively buoyant, rise quickly to be extruded along volcanic zones parallel the plate margin. These are the typical andesitic volcanics of Andean-style chains. Together these rock associations are typical of orogenic belts throughout the geological record.

While the principle that orogeny accompanies collision is not in question, the reality, as revealed in the rock record of the continents, suggests that things

units may move more-or-less intact for considerable distances. Such mélange zones are extremely complex, and the complexity is compounded by the metamorphism they experience as they descend quickly

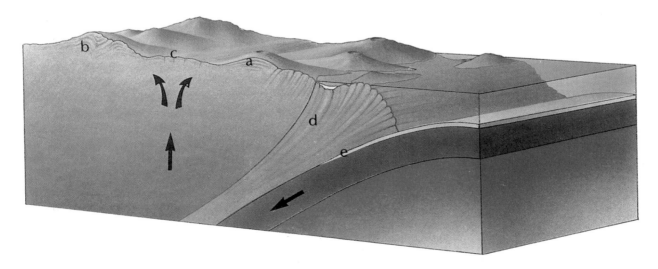

Diagram showing the plate situation along a convergent plate margin where an island arc has formed. A fore-arc (a) lies closest to the subduction trench. This consists of deformed oceanic sediments and is volcanically active. The back-arc (b) lies closer to the continental margin and is not volcanically active. The two are separated by a marginal basin (c) which develops as plate motion gives rise to rifting of the crust on which it is sited. On the landward side of the trench is what is called an accretionary prism (d) formed from the oceanic sediments that have been scraped off the descending oceanic plate (e).

(a)

(b)

Mélange deposits of late Pre-Cambrian age, Lleyn Peninsula, North Wales. (a) The house-sized white blocks represent a bed of quartzite that has been disrupted and which has slid down the ancient sea-floor. (b) These blocks are set in a finer matrix of similarly disrupted quartzite beds.

are not always as straightforward as they seem. In particular, it often seems that the first in a series of orogenic episodes pre-dates the first recognizable continent–continent collision. To endeavour to elucidate this, first we must take a closer look at various collision scenarios.

Firstly let us take the situation where a plate with continental crust on its leading edge is converging upon another with oceanic crust on its leading edge but carrying continental crust that still lies at some distance from the collision zone. Initially the oceanic

lithosphere will be subducted beneath the plate edged with continent, a deep trench having formed within the oceanic crust oceanward of the contact zone. Eventually, when all of the ocean crust is subducted, subduction will cease along the earlier destruction zone, and subduction will likely begin in the opposite direction, with generation of a new line of volcanoes along the opposite plate margin. The descending oceanic slab will gradually become detached and sink into the mantle, there to be melted and recycled – a process which removes the downward pull on the plate margin – therefore the overlying crust rises up. A new mountain chain may arise parallel to the collision zone. So far, so good.

We now need to take take a step backward and have a slightly closer look at a typical subduction zone, at a stage when continental crust is still not involved. Continentward from the trench, where the plates are in contact, what is called a *fore-arc* forms.

127

This is constructed of scraped-off slivers of oceanic crust and sea-floor sediment (including mélange) which are thrust one above the other until a point is reached when the arc may grow above sea level. Before it reaches this stage the rocks will have become strongly deformed and metamorphosed, so that any islands which eventually emerge will be built from a complex pile of tectonized materials. Geologists call this a *sedimentary arc.*

As soon as the island chain appears above sea level, erosion takes over, debris being deposited on either side of the island chain, much of it in the trench (modern Barbados is an example). Behind the sedimentary arc a *fore-arc* basin will form, this receiving sediment too. Behind it grows a *volcanic arc*, this rising above the subducted slab where that has reached depths sufficient for its melting to take place. The volcanic arc will consist of voluminous andesitic lavas and pyroclastic rocks, the latter being particularly abundant. Island arc volcanoes are usually dangerous, being unpredictable and violent in eruption. The Indonesian volcano, Gunung Merapi is one such example; Krakatoa is another. Finally, a *back-arc basin* may lie on the far side of the subduction region.

Where an ocean–continent collision is occurring the sequence of events will be similar; however, the deformed sedimentary arc will then lie directly against the edge of the colliding continent, a chain of islands therefore will not grow up, while the volcanic arc becomes a land-situated structure (the Andean coast of South America is a modern example). Thirdly, should the scenario be extended to include the subduction of a spreading axis, then once the oceanic ridge becomes subducted, spreading will terminate and the preceeding part of the oceanic slab will become detached and sink into the mantle. This leaves a residual slab of oceanic crust which, assuming convergence continues until the two continents collide, will eventually be thrust onto the preceeding edge of the continent, forming an ophiolite sequence. Thus it is that ophiolites are evidence for former oceans.

Reverting now to the business of orogenesis: future mountain chains are built from the deposits of trenches, fore-arc basins and volcanic arcs involved in collision zones. However, when we look closely at the geological record, it seems – as I mentioned earlier – that much fold-mountain building precedes the closing of oceans and associated major continental collisions. Yet there seems little likelihood that the huge fold-mountain chains raised during the various major orogenies could have developed purely from subduction of or collision with island arcs. In consequence some geologists have suggested that large slabs of thickened oceanic or continental crust may have lurked offshore from continental margins, and that it was collisions with these which may have triggered off many orogenic episodes. There is some evidence that such 'rafts' (micro-continents?) of anomalous crust may have existed, for instance, one such has been discovered in British Columbia and southeast Alaska, another in Oregon. Currently geologists are seeking further such examples in a bid to explain fully the observed stratigraphic and structural relationships within the framework of plate tectonic theory.

14

Late Pre-Cambrian and Early Palaeozoic times

Introduction

Toward the close of Proterozoic times the Earth experienced a prolonged period of crustal instability, during which a worldwide series of mobile belts developed. Most of these remained active until well after the beginning of the Palaeozoic era. In Laurasia, mobile zones formed around a number of small cratons, many of which were to restabilize before the Lower Palaeozoic was out. Activity within this series of belts spans the period between 900 and 370 million years ago.

In North America an orogeny occurred about 470 million years ago; this was the Taconic orogeny. Prior to this we believe a subduction trench existed on the site of eastern North America, receiving sediments from a landmass to the west. This accumulated on the site of the Appalachians. The orogeny was accompanied by the usual geological phenomena: volcanism, folding and the uplift of a landmass, in this case to the east of the present Atlantic seaboard. However, this upheaval closed no ocean; this was to come much later. Indeed, subduction, volcanism and sedimentation continued until around 370 million years ago, whereupon the northeast tip of North America collided with a part of what is now northern Europe. The orogeny which resulted from this collision has been named the Caledonian–Acadian orogeny and it produced a huge mountain chain, the remnants of which are to be seen in Scotland, Norway, New England and eastern Canada. Gradually the ancient ocean – the Iapetus Ocean – which separated the northern continents was closed, but not until around 300 million years BP.

The Caledonian mobile belt

The term 'Caledonian' was first introduced by the Austrian geologist Eduard Suess (1831–1914) who applied it to a belt of strongly-folded rocks found in Great Britain and Scandinavia, which were overlain

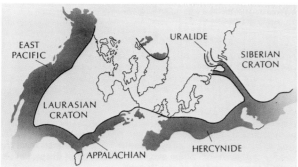

The mobile belts that formed along the margins of Laurasia during mid-Palaeozoic times.

by deposits of Old Red Sandstone (Devonian) age. The cycle of events that produced these rocks culminated in the Caledonian orogeny, the rocks and the orogeny also being known as the 'Caledonides'. Its effects were felt well beyond the European region, extending into North America and beyond (here known as the Acadian orogeny). The earliest recognizable Caledonian events date from around 800 million years BP, when broad basins begun to receive considerable volumes of sediment. Before the close of the cycle these had accumulated an amazing 15 km of material.

The active zone flanked the edges of both the Baltic Shield and the European craton. Both of these ancient blocks still remain intact. On its west it flanked an ancient craton long since fragmented; however, a tiny remnant is to be found in northwest Scotland, but the largest pieces are in central and western Greenland.

The cycle of events

The Caledonian cycle began, as I have said, with the formation of sediment-accumulating basins, formed as downwarps in ancient cratonic blocks. Some formed early in the cycle, while others evolved later on. Orogenic activity began in the earlier basins even before sedimentation had ceased and it continued right through into Palaeozoic time. The orogeny produced metamorphism and deformation of the sedimentary pile. Those rocks which collected in the older basins suffered quite intense changes, technically called *high-grade metamorphism*; those of the younger basins escaped more lightly and were metamorphosed to a lower grade.

About 400 million years ago the final phase of the orogeny began: the whole region was gradually uplifted into a majestic new fold-mountain chain – the Caledonian Mountains. These would have stood proudly above the continental landmasses for millions of years, but eventually were worn down by the inexorable power of erosion, the resultant debris being spread out at the foot of the new mountains, or in new basins formed between adjacent mountain massifs. The new breed of sediments belong to the Old Red Sandstone (Devonian) and were spread out on continental crust under relatively arid climatic conditions – very different from their marine Caledonian predecessors.

Subsequently, compression of the belt – due to plate collisions – led to much fracturing of the rocks which were invaded by granitic magma. By about the middle of Devonian times (350 million years BP), the mobile belt had stabilized in northern Europe and in the northern Atlantic region, but stabilization was delayed in the more southerly regions of Europe and in North America. In the latter regions the mobile belt was transformed into another great mountain chain in Late Palaeozoic times.

Sedimentary basins – generalities

Sedimentary basins are one of the most characteristic features of the Earth's lithosphere, with a range in age that spans the Archaean to the present time. Currently the largest basins are located at the margins of Earth's continents, for instance, near the mouths of major river systems such as the Amazon, where 15 km of sediment has accumulated. However, there are also large basins within the continental interiors, for instance the Russian Platform and the central North American basin of Illinois and Michigan. These are called *intra-cratonic basins*. Basins are also associated with foreland regions, i.e. the flanks of mobile belts (e.g. the Alps and

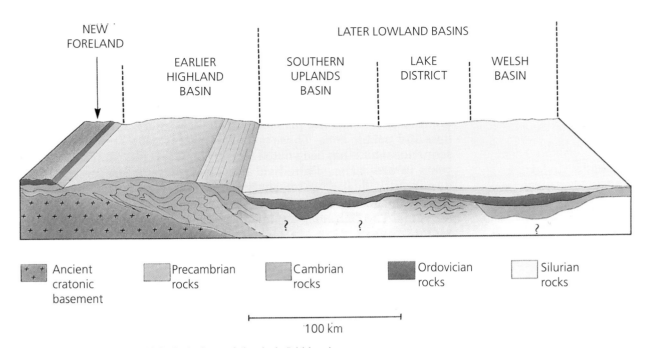

NEW
FORELAND

LATER LOWLAND BASINS

EARLIER
HIGHLAND
BASIN

SOUTHERN
UPLANDS
BASIN

LAKE
DISTRICT

WELSH
BASIN

| Ancient cratonic basement | Precambrian rocks | Cambrian rocks | Ordovician rocks | Silurian rocks |

100 km

Schematic diagram showing Caldedonian basin morphology in the British region.

Appalachians). The development of basins is dominated by vertical movements for most of their evolution, and it is only when they are later caught up in plate movements that lateral effects become important and basin-filling materials become tectonized. However, it has been recognized also that an initial depression is a prerequisite of basin instigation, this being provided by rifting, compression or strike-slip, i.e. deformation.

We still know little about the deep structure of most sedimentary basins, yet we do have a pretty good idea of what lies beneath one, the North Sea Basin of the Atlantic province. Within this basin there is a 9-km-deep pile of sediments ranging in age from the Palaeozoic to the present day. Seismic data indicate that beneath this the Moho – the Mohorovicic Discontinuity, a seismic boundary across which the velocity of seismic waves abruptly changes – lies at a depth of between 18 and 25 km while, beneath the adjacent areas of Great Britain and Scandinavia, it is deeper, at between 25 and 35 km. The Moho is thus shallower beneath the basin itself. Evidence for rifting in the subcrust is found here.

The most widely-quoted mechanism for basin formation is the thermal contraction which the oceanic

crust experiences as it moves away from a spreading centre. This may be added to by the overburden of unusually thick piles of sediment; for instance the turbidites which depress the Atlantic floor by at least 1 km more than expected, and the mouths of the Niger and Mississippi Rivers, where sediment derived from the adjacent continents depresses the oceanic lithosphere by as much as 4–5 km more than expected from the age of the crust alone. Generally, however, more than 2-km thickness of sediment is rare on oceanic crust.

The thickest sedimentary sequences are to be found on continental lithosphere, generally at or close to the present-day continental shelf break in slope. One particularly notable locality – off the eastern Newfoundland coast – boasts between 10 and 15 km of Mesozoic to Tertiary age sediments overlying a crystalline basement of Palaeozoic or Pre-Cambrian age. The great thicknesses found in such locations may well be due to the attenuation of the crust during early rifting, the thinned continental crust then cooling, subsiding and providing a suitable depression for continued sediment deposition.

Sedimentary basins – specifics

At the beginning of the Caledonian cycle sediments were laid down on an eroded land surface built from gneisses of Lewisian (Proterozoic) age. This represented a small fragment of an old craton that existed before continental drift split apart Laurasia, and the greater part of which now resides in Greenland. The earliest rocks, known as Moinian rocks, are a thick series of rather monotonous sandstones that appear to have built out into the sea as large deltas. They reached a maximum thickness of around 7 km, and have been radiometrically dated at between 1000 and 800 million years old.

On the western side of this belt, where the sediments were in direct contact with the cratonic basement, there is what is termed a *foreland* region. It was here that the deformed sediments were thrust over the cratonic basement later in the cycle, as two continents collided and exterminated the early ocean that once separated them. In the foreland region the sediments are thinner, consisting primarily of shallow-water sandstones, boulders and pebble beds. These belong to what is called the Torridonian sequence which has been dated to the same period as the Moinian. Somewhat later than both, the Dalradian rocks accumulated. These, predominantly deep-water shales, limestones and immature sandstones, were followed by glacial deposits – indicating a gradual change in the climate as time progressed. The Dalradian basin deposits reached a similar thickness to the Moinian.

In eastern Greenland and Spitzbergen there are

Late Pre-Cambrian/Lower Palaeozoic sedimentary sequence on the east coast of Greenland. In places the sediments are 200 km wide and include rocks which range in age from late Pre-Cambrian to Ordovician. [Photo: Jack Soper]

Folded and faulted granitic gneisses of Lewisian age, Lochailort, Scotland. These rocks are the ancient cratonic basement upon which the Caledonian rocks were laid down.

Folded Moinian sandstones and siltstones, Tarskavaig, Skye.

thick sequences of sandstones, mudstones, limestones and glacial beds (called *tillites*). In the former locality they occupy a broad zone 200 km wide, containing rocks that have suffered high-grade metamorphism. In Scandinavia, the older basins immediately adjacent to the ancient cratonic foreland also collected sediments which show a spread in age from Pre-Cambrian to Ordovician. Further from the orogenic front are younger basin-filling deposits including feldspathic sandstones and tillites; these are up to 6 km thick and represent shallow-water deposits.

During the later part of the Caledonian cycle there appear to have been basins only in southern Britain. Here a series of subsiding, elongated, troughs received substantial volumes of sediments that included volcanic debris and turbidite sandstones, both indicators of unstable tectonic environments. Presumably this collected along an unstable continental margin, i.e. close to an active plate boundary.

One of the characteristics of these Caledonian basin-filling sequences is the presence of the glacial rocks called tillites. These bouldery deposits usually are found interbedded with marine sediments and have been attributed to a glaciation called the Varangian Ice Age. We believe they were dropped

133

from floating sea ice. The glacial beds seem to occur stratigraphically below fossiliferous Cambrian strata and are presumably of much the same late Proterozoic age wherever they occur.

More about basins

The subsiding basins into which these enormous volumes of sediment collected must have been huge, and it may seem anachronistic that much of the sediment accumulated in shallow water. This fact was stumbled upon first by James Hall in the mid-nineteenth century. Hall, an American geologist based in New York State, noted that the deformed strata of the Appalachians were much thicker than those of the same age found in the American Mid-West. In the Appalachians the total thickness was of the order of 12.5 km, but on reaching the valley of the Mississippi, this had dwindled to 1.5 km. Hall suggested that the Appalachian strata had accumulated in a slowly-subsiding basin which had received debris over a very lengthy period. Some years later another American, James Dana, proposed that such troughs should be called *geosynclines*. This term remains in the literature but has suffered much misuse and, to avoid confusion, is not used in this book, unless absolutely unavoidable. Instead I shall refer to depositional or *sedimentary basins* and identify specific features where necessary. We now realize that basin formation and evolution is a very complex process which though not entirely understood, is of vital importance in the understanding of how the Earth's crust behaves.

During the late Proterozoic and early Palaeozoic, two main series of basins appear to have existed, each group consisting of a number of smaller ones. The rocks now found in more northwesterly locations are very similar to the strata found in the northern Appalachian basin, which is of similar age. Those towards the southeast more closely resemble strata located in eastern Newfoundland, Nova Scotia and New Brunswick.

To take the northwestern basins first: their remains are to be found in the very northwestern tip of Hebridean Scotland and the western part of Spitzbergen, where limestones and other carbonate rocks unconformably overlie Pre-Cambrian rocks. Apparently these were deposited along an ancient passive continental margin. Further southeast, on the other hand, are found deformed schists (altered shales) and amphibolites (altered basaltic lavas); these were marine sediments and volcanics which had accumulated in deeper water and now underlie a tract that runs from northwestern Ireland, through northwest Scotland and via the extreme western edge of Norway into east Spitzbergen. Although the majority of these rocks can be shown to rest on sialic crust, towards the southeastern margin of the belt there are indications that the basins lay on oceanic crust, i.e. on an active plate margin.

The southeastern basins were filled by shales, greywackes, sandstones and a variety of basaltic and andesitic rocks; numerous unconformities exist within the sequence, indicating periodic upheavals. This is the typical assemblage of an active plate margin undergoing subduction. Thus the volcanism and deformation which affected the region during Early Palaeozoic times may have been produced by the passage of a plate bearing ancestral North America beneath that containing ancestral Europe.

Continental shelves

Sediments which are deposited under the relatively quiet conditions typical of a passive continental margin – akin to the present North Sea – gradually emerge as orderly sequences in the large basins. In time, wedge-shaped accumulations of sediment form from debris won by erosion from the adjacent continent. This spreads out as a broad nearshore shelf which extends oceanward to be truncated by the steeper slope of the continental slope which, in turn,

Lower Cambrian turbidite sandstones, Hell's Mouth Bay, North Wales. The thicker beds are about 2 m thick. Each unit is the product of rapid deposition from a turbidity current which flowed down the flanks of a Caledonian basin during early Palaeozoic times. The incut, thin bands are shales, representative of the normal deep-sea sediments into which the turbidites flowed.

descends to the abyssal plain.

Where rifting succeeds in breaking apart a continental block, as new oceanic lithosphere cools and contracts after it has emerged from the spreading axis, so the trailing edge of the attached continent slowly subsides due to inceasing density. In so doing, it allows the ocean basin to collect land-derived debris; as a result the sedimentary burden depresses the adjacent crust still further, enabling the basin to collect more and more sediment. Calculations indicate that for each 2 m of crustal subsidence, about 3 m thickness of sediment can accumulate. In this way accumulations may greatly exceed 10 km in thickness.

Beneath a wedge-shaped mass of basin sediment it is usual to find rifts incised into the subcrust; such would have opened in response to early tensional stresses within continental crust prior to separation. Basaltic lavas and continental deposits would have built up on the rift floors, the latter probably in lakes, with further deposition on the exterior flanks of the rifted dome. At a later stage, as the true oceanic crust evolved, slopes would reverse, with most sedimentation being towards the growing ocean. This is the stage at which sandy material would begin to be dumped on the continental slope and, during seismic activity associated with early instabilities, would be shaken down the slope onto the deeper ocean floor. This would be accomplished by turbidity currents – turbid suspensions of water, sand, mud and volcanic debris – which generate *turbidite sandstones*, very immature, poorly-sorted deposits. These turbidites would become interbedded with the normal, fine-grained deeper water sediments of the abyssal plain. Gradually a great thickness of mixed sediment would accumulate on a slowly-subsiding basement.

Later in the life of such a basin, as a shelf of sediment gradually builds outwards from the continental margin, sedimentation becomes dominated by finer-grained shales and limestones, as the supply of land-derived material dwindles. This occurs as the level of the land is lowered by a lengthy period of erosion.

Schematic section through the foreland region of the Caledonian mobile belt. Flat-lying shelf sediments typify the region northwest of the main thrusting. To the east these may be involved in thrust faulting. Further east again, large recumbent folds typify the Moinian and Dalradian rocks.

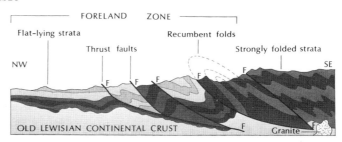

Such a pattern is typical of geological successions seen all over the world, and is an integral part of each geological cycle.

Continental forelands

When we look along the western margin of the Caledonian mobile belt – represented now by the rocks found in northwestern Britain and Greenland – we find that there are no strata younger than mid-Ordovician (around 460 million years BP). In northern Greenland, however, the old craton is covered by a sequence of limestones and shales that are fossiliferous and can be shown to span the time interval: Cambrian to late Silurian (570–430 million years BP); similar rocks are found also in northern Canada. Their presence illustrates that shallow seas must have spread over large regions of the adjacent craton during this period.

On the opposite side of the mobile belt, i.e. in the region now respresented in the rocks of parts of Scandinavia and the Shropshire region of Great Britain, Lower Palaeozoic shallow-water sandstones and other shoreline deposits also bear witness to the presence here of a shallow sea. Presumably the well-known English viewpoints of the Longmynd and the Wrekin must have risen from the waters of this ancient sea, as islands. In Norway, particularly around Oslo, and in southern Sweden, shallow-water sandstones, limestones and shales of similar age are found. By later Palaeozoic times, however, the foreland regions gradually emerged from beneath the shallow ocean, as the terminal events of the geological cycle wreaked their havoc.

Away from the foreland regions, in the core of the mobile belt, tremendous upheavals rucked up the sedimentary strata into large folds, many of which became recumbent, i.e. they were forced over on their sides, so as to rest horizontally. Tremendous tectonic forces – associated with the collision of plates – drove many of these folds outward from the core regions, so that they were thrust over the foreland zones, coming to rest on the ancient cratonic basement made of gneisses and metamorphosed sediments.

Furthermore, metamorphosed sediments derived from the core regions were sent out as low-angled thrust sheets in the same direction. Such thrusts are typical features of foreland zones, and are typified by the great Moine Thrust of northwestern Scotland. This major zone of lateral displacement saw great slivers of rock translated many tens of kilometres during the terminal events of the cycle. Radiometric dating of the rocks altered by these events indicates that the last movements along the Thrust took place around 430 million years ago.

Folding and metamorphism

One of the characteristic features of sedimentary and, indeed, many volcanic rocks, is that they are horizontal or subhorizontal when first laid down. It only takes a brief look to see that this is not the case for the rocks found in orogenic belts, be they sedimentary or volcanic. These very active regions of the Earth are typified by rocks which have been rucked up into folds varying in scale from tiny microfolds to huge anticlines and synclines measurable in tens of kilometres. Many are cut by thrusts and other types of fault.

Of course, not every rock type responds to deformation in the same way; a brittle sandstone or limestone takes up strain in very different ways from, say, a shale. The precise manner in which a material reacts to pressure is dependent upon internal properties such as chemistry, strength and the amount of water contained in pore space, and upon external factors such as the pressure and temperature under which it is confined and the rate at which compres-

Folded and faulted Cambrian sandstones and siltstones (now converted to low-grade schists), Porth Ceiriad, Lleyn Peninsula, North Wales.

sion is occurring at any given time. The same rock, confined under different conditions, will respond in quite different ways to the stresses imposed on it.

Regardless of the exact conditions, the general outcome of deformation is the generation of folding or, if the rocks are brittle, of fractures and faults. The San Andreas Fault is perhaps the world's most notorious fault, if only because it runs through some of the most densely-populated parts of California. However, it is only one of millions of similar structures that can be identified within the crust of the Earth. The same is true of folds. Every episode of deformation produces its own particular fold regime, the Caledonian orogeny being no exception. During the long span of time during which this mobile belt evolved, the rocks within it were often deformed several times, as many as five or six in some regions. As a result folds, refolded folds and refolded refolded folds are commonplace. The geology can become exceedingly complex! One of the tasks of the trained structural geologist is to unravel the folding complexities, trying to unfold the strata and identify the palaeogeography prior to deformation.

Naturally, deformation takes place inside the Earth, not at the surface, thus it occurs under conditions of elevated temperature and pressure. Since sedimentary rocks are formed at surface temperatures and pressures – their constituent minerals being sta-

ble under these conditions – it is not surprising that they become rather uncomfortable at depth. The result is that their atomic lattices rearrange themselves to accommodate the new environment, giving rise to new minerals that are stable under the new set of conditions. This process of change is called metamorphism; it takes place on the atomic scale, in the solid state, and does not involve actual rock melting (at least not until the conditions are unusually severe).

One of the first geologists to study metamorphism within an orogenic belt was George Barrow who, during the latter part of the nineteenth century, studied the Dalradian rocks in the Grampian Highlands of Scotland – a part of the Caledonian mobile belt. He noted that as he crossed the Grampian Mountains from southwest to northeast, a succession of different assemblages of 'index' minerals appeared in the metamorphic rocks. These minerals, produced in response to regional metamorphism of the Dalradian sedimentary and volcanic rocks during the

Tightly-folded Lewisian gneisses, Ardgour, Scotland.

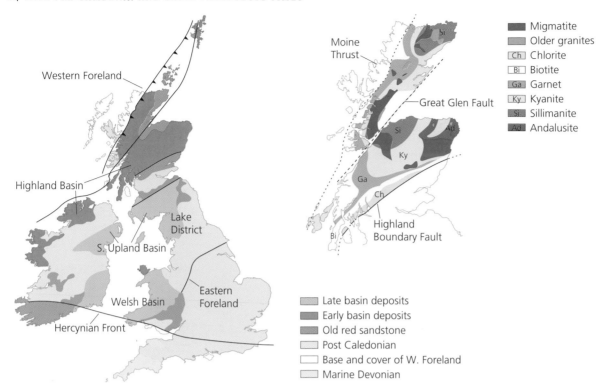

The Caledonian mobile belt as developed in Britain: (a) the main structural elements of the belt; (b) an enlarged map of the Highland zone, showing the arrangement of index minerals produced during Caledonian burial and metamorphism. (Ch=chlorite; Bi=biotite; Ga=garnet; Ky=kyanite; Si=sillimanite; Ad=andalusite.)

Caledonian orogeny, apparently defined zones of progressively increasing grade of metamorphism, the most intensely altered rocks outcropping towards the northeast.

In mapping the region Barrow had stumbled across the fact that during metamorphism a thermal gradient is established, the highest grade rocks representing the core regions of the orogenic zone. Thus it was that, in moving westward away from this core, rocks of lower and lower grade were encountered, the lowest lying near to the foreland zone. This, of course, would have lain near the margins of the orogen. A similar pattern has since been detected in most such belts, although the specific index minerals vary, depending largely on the chemical composition of the original rocks involved in the deformation and metamorphism.

Caledonian magmatism

As is the case with any orogenic episode, magmatism was another important ingredient of the geology.

Much of this was submarine in nature, with the result that pillow lavas are commonplace in the Lower Palaeozoic rocks of Wales and southwest Scotland; however, there were also significant volcanic loci along the continental margins and some of these may have been island arcs. Basaltic lavas are abundant within the marine sedimentary succession, but associated with the marginal successions are andesites and rhyolites. The latter were particularly important during Ordovician times.

One of the most significant events, however, was the rise of huge volumes of granitic magma into the crust during the latter stages of the cycle. There are at least fifty separate granite bodies in Britain alone; about one tenth pre-date 450 million years BP, the rest are younger. Similar intrusions are found in Scandianavia, Greenland and Spitzbergen. Radiometric dating indicates a range of ages between 600 and 390 million years, but a significantly large proportion of the plutons were emplaced around 400 million years ago, during what was probably the final thermal event during the Caledonian cycle.

Caledonian granite, showing prominent pink K-feldspar crystals, together with black mica and greyish quartz.

In the central regions of the Caledonides, the granites are associated with *migmatites*, rocks with mixed affinities, being at the same time magmatic and metamorphic; they are characteristic of the deeper 'root' zones of orogens. Elsewhere granites have thermally metamorphosed the country rocks, indicating that the latter were relatively cool when intrusion occurred. Accompanying most granites are *pegmatites* which typically occur as veins and blotches. Such rocks are the products of the later stages of crystallization and as a result are enriched in volatile constituents, many containing unusual minerals that accommodate the less common elements, i.e. those which do not enter the lattices of the more common rock-forming mineral species. Because they are volatile-rich, the crystal size in some pegmatite bodies is strikingly large, giving rise to rather spectacular rocks.

Not all of the granitic bodies formed in the same way; some evidently were emplaced forcefully, pushing aside the crustal rocks as they rose. Others have a much more complex emplacement history, having risen into the crust as a number of separate pulses; one such example is well known to British geologists, the complex of Glen Coe (Scotland). The emplacement involved the subsidence of a cauldron-shaped block of crust which enabled the magma to enter and solidify.

Photograph showing the roof region of a light-coloured granitic intrusion where veins and bosses of the granite have invaded the country rocks above. Bonawe, Scotland.

The final stages

Towards the close of Silurian times, about 395 million years ago, a number of important changes in the palaeogeography of both Europe and the North Atlantic had come to pass. Firstly, temperatures within the crust had fallen appreciably within the mobile zones. This is shown by the abundance of K–Ar radiometric ages which cluster around 410–450 million years BP. The fact that such ages have registered shows that the orogen had cooled sufficiently for argon to be retained in the crustal rocks, setting the radiometric clock. Secondly, most of the folding appears to have ceased, while, thirdly, changes took place in the pattern of sedimentation, new basins being formed on the repositioned continental crust. These new basins were to receive a great thickness of sediment, won from the newly-risen landmasses by rapid erosion of the Caledonian mountains. As the mountains were worn down so, according to isostacy theory, the land rose in response to achieve isostatic equilibrium. Furthermore, large volumes of granite having been emplaced between 410 and 380 million years ago, the continent became even more buoyant. In this relatively short period – about one tenth the length of the complete cycle – a great deal of activity had been crammed in.

This, then, was the close of the Caledonian cycle. It was marked by rapid erosion of new mountain chains which rose isostatically to compensate for this. This resulted in the accumulation of thick continental deposits traditionally known as the Old Red Sandstone. As would be expected, these were thickest towards the margins of the old mobile belt, which marked the lower flanks of the newly-formed mountains. Over 8 km of such strata were laid down in eastern Greenland at this time. The Old Red Sandstone is, however, part of another story. Suffice it so say that the change in palaeogeography had a profound effect on the global climate.

Pre-Cambrian ice ages

As we have seen, the Caledonian cycle spanned the period between the late Protorozoic and the end of the Early Palaeozoic era. It provides a link between the lengthy period when geological data are gleaned with some difficulty, and not without a measure of uncertainty, and more recent events when fossils became abundant and the details of stratigraphy become clearer. Despite the difficulties associated with the older rocks, sufficient evidence has been forthcoming to establish that glaciations occurred very early in our planet's history.

There is no definite evidence that an Archaean ice age occurred. This may in part be due to the absence of very large landmasses, or in part to unsuitable climatic conditions, or both. The situation is rather different for the Proterozoic, however, there being little doubt that ice sheets covered large regions of the North American craton several times. Furthermore, supposed Proterozoic glacial deposits have been reported from all of the continents apart from South America, and it appears that the late Proterozoic glaciation was one of the largest to have occurred, its deposits reaching unusually low latitudes. Exactly why ice ages occur is something of a mystery; the topic will be discussed more fully in Chapter 24.

The deposits which represent the glacial conditions are the bouldery beds called tillites, together with rhythmically-banded sediments akin to varved deposits of more recent ice ages, and 'dropstones' – boulders that appear to have been dropped from drifting ice into submarine sediments. Together these provide pretty convincing evidence for very cold conditions.

Some of the most convincing evidence is to be found amongst the early Proterozoic strata of the region around Lake Huron, Canada. During the period in which 12 km of sedimentary strata were laid down, there is evidence for at least three glaciations. The youngest telltale horizon, called the

Late Pre-Cambrian varved sediments, Ella Island, Greenland. The banding is the result of the glacier's seasonal cycle of freeze-up and melt, which releases pulses of sediment during the warm season. The coarser sandy bands were deposited quickly but the intervening white silt bands and clays took much longer.

around 600 million years ago. Glacial horizons associated with these events have been located on all of the continents bar South America.

Exactly why glaciation occurred at these times is a matter of considerable debate; some scientists believe it to be due to drifting of early continents through the polar regions during the very first widespread phase of continental drift. Others cite a rather special type of mountain building associated with drift, while others see an explanation in an 'antigreenhouse' effect, where carbon dioxide was lost from the atmosphere just prior to glaciation. It is probably correct to say that at present no one explanation is universally accepted.

Blueprint for a geological cycle

The geological processes which accompanied the development of the Caledonides were neither confined to that particular period of time, nor affected only the regions decribed. Similar events have affected the lithosphere elsewhere, at both the same and different times. I have chosen to introduce Caledonian geology simply as it has been widely studied and provides us with a blueprint for other cycles.

Continent-sized portions of the Earth's crust whose rocks, fossils and geophysical evidence imply that once whey were situated at quite different latitudes, can only be interpreted as meaning they have drifted around. Similarities between cratons now widely separated, perhaps by ocean, perhaps by belts of younger rocks, allow us to piece together the fragmentary evidence and show how continental collisions and splittings apart have shaped and reshaped the Earth's palaeogeography. It is now time to take a look at other regions and at other mobile belts, in order to see how the Earth developed and in so doing, extend our palaeogeographical knowledge.

Gowganda Formation, is very extensive and covers an area in excess of 120 000 km². It is considered to be about 2300 million years old. Similar rocks are found elsewhere in Canada and detailed studies suggest that during this period sediment was being deposited in a radial pattern, as though laid down around the edges of a continental ice sheet.

A search for similar horizons has been made on other continents too. There is good evidence among the Griqualand tillites of southeast Africa, and the Turee Creek Formation of northwestern Australia, for a glaciation at much the same time. Palaeomagnetic evidence, admittedly rather sketchy, suggests that the centre of glaciation was probably around a southern palaeolatitude 60°.

Later in the Proterozoic, between 950 and 570 million years BP, there appears to have been even more widespread glaciation. A first period spanned the interval 950–900 million years BP, a second the period around 750 million years BP, and a third

The largest monolith in the world: measuring 8.8 km around its base, and 348 metres high, Ayers Rock (Uluru to the Aboriginals) is one of the greatest single natural attractions in Australia. Composed of tilted red sandstones and conglomerate, it is a remnant of ancient Pre-Cambrian strata that underlie vast areas of the Australian craton. In the distance are the rounded massifs of the Olgas (Kata Tjuta), which are made of very coarse conglomerates of similar age.

15

The Appalachian story

Introduction

North America in Proterozoic times was very different from the continent we know today. As we have seen, the ancestral continent did exist and at times formed a part of Laurasia, one of the early supercontinents; however, its core was quite small at this time. During the later part of Proterozoic times the cratonic cores of ancestral North America, Europe and Gondwanaland were separated; indeed, they were drifting apart. Later, during the early Palaeozoic, thick deposits of sedimentary and volcanic rocks accumulated in elongate basins which developed along the continental margins of ancient North America. The plate movements apparently were then reversed, so that eventually the sediment-filled basins of the continental margins became crumpled up to form a part of a worldwide system of orogenic belts, of which the Caledonides were just a part. Those which grew along the North American margins are called Appalachian belts, and include the Ouachita, Cordilleran, Flanklinian and East Greenland mobile belts. They did not all form at the same time, and the end of this great cycle was only reached when the final vestiges of the ancestral Atlantic – called the Iapetus Ocean – disappeared down a long gone subduction zone.

Iapetus – precursor of the Atlantic

As with all orogenic episodes, the Taconic upheavals which affected the northern Appalachians were related to plate movements. Thus, in the latter part of Pre-Cambrian times – between 800 and 700 million years BP – ancestral North America split away from Gondwanaland, specifically the part of it now represented by Africa. This led to the opening of a new ocean called Iapetus. Ancient spreading centres must have bisected it, just as they do in the modern Atlantic, of which it was the predecessor. Iapetus widened for a lengthy period but then, around 600 million years ago, during the final stages of the Pre-Cambrian era, it began to close again. Eventually this brought about the convergence of plates during the Cambrian until eventually folding, faulting and volcanicity left their indelible marks in eastern North America.

It is thought that during the Late Proterozoic a substantial fragment of continental crust became separated from the main American craton, a marginal sea developing over the oceanic lithosphere between the two. On the receding margins of this 'raft' and mainland America, shallow-water shelf sediments accumulated. In the early Cambrian the Iapetus Ocean began to close and continued so to do until the early Silurian – a time interval of about 50 million years. During the early stages there was subduction adjacent to the plate margins and probably island arcs lay offshore. Subsequently the estranged continental fragment crashed into North America, with much folding, thrusting and folding of the rocks along the collision zone. This produced the Taconic orogeny in northern Appalachia.

Eventually, between 500 and 400 million years ago, the island arc also was pushed against North America, producing a second phase of folding and thrusting which was even more intense. This event is known as the Alleghenian orogeny and it rucked up the southern part of the Appalachian mobile zone. It also saw the closing of the more southerly parts of Iapetus.

It is calculated that the lateral movements produced by the combined orogenies pushed some of the thrust sheets at least 250 km westwards over the margins of ancestral North America. Iapetus was finally eliminated completely during Late Carboniferous–Early Permian times, in what is called the Appalachian orogeny. This occurred when the South American corner of Gondwanaland ran into Texas and Oklahoma, raising up the Ouachita and the Marathon Mountains. The southern Appalachians

were then thrust up by a further collision this time with northwest Africa. This fragment is now the state of Florida – a massive limestone plateau.

The continental interior

In excess of 5 km of early Palaeozoic marine sedimentary rocks accumulated along the eastern margin of North America. Calculations indicate that in the strip between Newfoundland and north Alabama, the deposits accumulated at a rate of about 13 m per million years. Compared to this, the rate of accumulation on the continental interior was very modest: only 1.5 km of sediment was laid down in the same period. It can be safely assumed that the continent was a low-lying landmass at this time. Moving towards what were the ancient mobile zones, things change quite dramatically: the thickness of late Pre-Cambrian sedimentary deposits increases. In places the oldest Cambrian strata lie unconformably on the uppermost Pre-Cambrian beds. The facies variations shown by the sediments, together with the current directions revealed by sedimentary structures preserved within the rocks and on the bedding surfaces, indicate that the source of the material lay in the continental interior.

The sediments themselves are what are known as 'mature', i.e. they consist of only resistant minerals (such as quartz, apatite, zircon etc). This implies that the less resilient minerals (such as feldspar, mica, amphibole) must have been winnowed out of the sediment inventory by a lengthy period of weathering and erosion that lowered the ancient continent to a low-lying region, and removed the non-resilient grains. The smoothed, sifted and sorted grains were then swept toward the continental perimeter by rivers and streams, there to build the relatively thick beds of clastic sediments now found there. Among the Late Proterozoic strata are tillites of comparable age to those found at the same level in Eurasia:

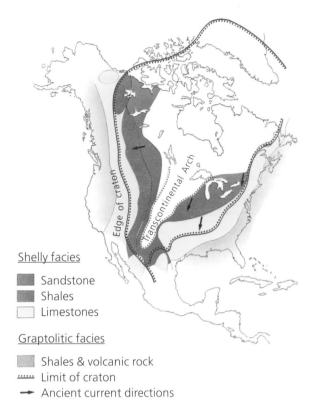

Ancient North America during the late Pre-Cambrian. The margins of the craton are indicated by the dark and dotted line. The main zones of deposition lay towards the edge of the continent, where there was an outward passage from shelly to graptolitic facies. The presence of sandstones, shales and some limestones inside of the cratonic boundary line shows how the sea transgressed over the low-lying marginal regions of the continent during late Cambrian times. The Transcontinental Arch apparently separated two regions of shallow-water deposition and began to rise during the Early Cambrian.

Shelly facies

- Sandstone
- Shales
- Limestones

Graptolitic facies

- Shales & volcanic rock
- .... Limit of craton
- → Ancient current directions

By late Cambrian times almost all of the land seems to have gone, the source of clastic sediment having been submerged. Above the predominantly clastic rocks, therefore, are found marine carbonates, like those currently being laid down in the vicinity of the Bahamas Bank. Most of the beds are composed of the broken shells of dead organisms, the kind of animals which live in well-oxygenated, agitated, shallow seawater. Such conditions are found today most widely in the Tropics and Subtropics, suggesting that North America enjoyed such a climate at that time. This is supported by palaeomagnetic data.

clearly glaciation affected North America too.

Somewhat later, during early Cambrian times, a broad arch grew in the continental crust; this extended from Arizona into the region of the present Great Lakes. The arch appears to have separated regions on either side where marine deposition occurred, a shallow sea extending over the craton during Cambrian times, covering vast areas by the end of that period. This Cambrian *marine transgression*, recorded by unusually widespread sedimentary deposits, was one of the most widespread geological events of the time.

The Cambrian strata are richly fossiliferous, containing trilobites, brachiopods and cephalopods – all creatures which lived under the sea (confirming the marine transgression notion). In all likelihood the craton rose and fell epeirogenically several times throughout this period, so that positive regions became separated from one another by depositional basins, the position of which altered as time passed. The arched, positive regions would have supplied sedimentary debris to the intervening basins, and to the Cambrian ocean.

Outcrop of Palaeozoic siltstones, west of Jerome, central Arizona. This region was arched up during Cambrian times, separating depositional basins from one another.

The Appalachian mobile belt

The Appalachian mobile belt developed along the boundary between ancient North America, Europe and Gondwanaland. It was the site of several elongated marginal basins in which accumulated a variety of rocks prior to convergence of the continents during the early Palaeozoic. When the intervening ocean was subducted and the preceding continental edges came together, the basins and their contained sediments were crushed and a new fold-mountain chain emerged where once there had been ocean. Today the eroded stumps of this great mountain barrier are found in a belt of deformed and metamorphosed rocks that extends from Newfoundland to Alabama.

Thick accumulations of sedimentary rocks typify the belt. At the start of Cambrian times marine sedi-

mentation occurred along the continental margins, but as time went by the sea encroached westward across the continental interior, covering the shallower basins. Little deformation appears to have taken place until the middle of Ordovician times, when the Taconic orogeny caused major upheavals. In west Newfoundland sheets of oceanic lithosphere – ophiolites – were thrust onto the continental margin around 510 million years ago. In all probability a subduction zone sloped down westward beneath the Appalachian mobile belt, while at the same time massive slides of mélange slipped down the continental slope into the trench. These deposits have Middle Ordovician beds both below and above them, fixing their precise age rather neatly. However, an ocean still existed east of much of the Appalachian region (Iapetus) and geologists have appealed to some mave-

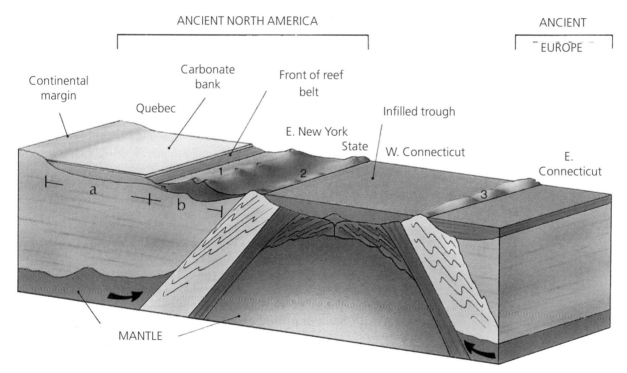

One possible reconstruction of the geology of the northern Appalachian mobile belt during Cambrian times. The infilled trough – now underlying western Connecticut – was formerly the site of the Iapetus Ocean. 1 = inner volcanic arc; 2 and 3 = outer tectonic arcs; a = Inner Appalachian Basin; b = Outer Appalachian Basin.

(a)

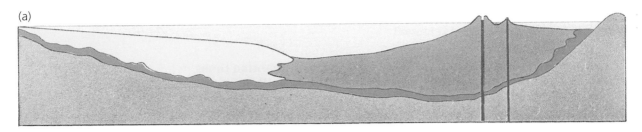

(b)

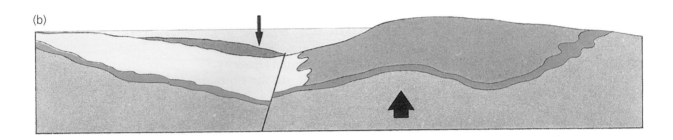

(c)

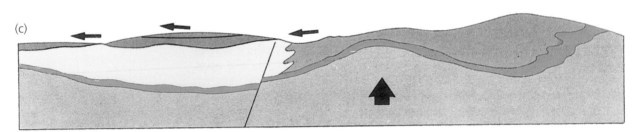

The production of laterally-moved slabs of crust (allochthons) in the region to the left of the previous figure, during the Taconic orogeny. (a) Deposition of basin sediments during Cambrian times; carbonate shelf deposits in the nearshore environment; deeper-water sedimentation further out; (b) uplift of the outer basin gives rise deposition of clastic sediments on top of the carbonate sequence; (c) gravity sliding of uplifted outer-basin sediments (allochthon) on to the inner-arc area.

rick mini-continent – lingering conveniently offshore – to have provided a source of collision that allowed the orogeny to raise up a new mountain chain from the trench sediments!

Whatever turns out to be the true explanation (and much work still remains to be done here), the uplifted rocks were rapidly eroded during Silurian times, as the ocean attacked the continent's borders. In central Maine, parts of Quebec and central Newfoundland there is evidence for a later phase of folding, accompanied by metamorphism and magmatic intrusion during the Late Silurian and early Devonian times. Radiometric dating yields 395-million year ages for the granites intruded into the mountain roots at this time. This corresponds roughly with the last events of the Caledonian orogeny in Europe.

The North American interior

During the Late Palaeozoic most of the craton was covered by a shallow sea. The more rapidly-subsiding basins attracted substantial volumes of sediment which presumably were derived from hilly regions between them. There is certainly evidence of uplift and increasing extent of highlands in the east and south of the craton during the Late Carboniferous, as increasingly large amounts of clastic debris were spread out in the eastern, central and southern parts of the interior.

The sea, however, did not remain at the same relative level throughout this time: there were repeated incursions and regressions over the almost flat continental interior. This gave rise to a distinctive, rhythmic pattern of sedimentation in which shale, sandstone and limestone repeated themselves in sequence, time and again. This *cyclothemic* deposition is the same as that which produced the extensive coal deposits of Carboniferous times in Europe.

The sea finally withdrew from the eastern side of the craton during the Permian period, whereupon coastal lagoons developed in what is now Oklahoma and Kansas. The large Delaware basin formed among the hills of the Marathon Mountains of northern Texas and southeast New Mexico. The deeper parts of this basin became flanked with reefs in Middle and Late Permian times, while limestones and dolomites accumulated in the shallow lagoons behind them. The lovely Carlsbad Caverns have been hewn out of these carbonate reef deposits. Palaeomagnetic data show that the basin was at a latitude of around 15° N at the time the reefs were formed.

Little deformation affected most of the interior, but during the Late Carboniferous, the Arbuckle, Wichita and Amarillo Mountains were raised up in Texas, while the ancestral Rockies began to rise in Utah, New Mexico and Colorado. Very thick accumulations of clastic rocks were deposited in basins adjacent to the flanks of these new ranges. We shall return to this story later.

Present-day features of the Appalachians

If we look at the present-day geological features of the southern Appalachians we perceive four distinct zones, each with their own particular characteristics. These developed as a direct result of their varying positions in relation to ancient plate margins. In the Valley and Ridge Province, thick Palaeozoic sedimentary rocks are found. These are intensely folded and show clear evidence of having been thrust northwestward by orogenic forces from the southeast. The tectonism they suffered came in three pulses, during the late Ordovician, at the end of the Devonian, and a third in the Permo-Carboniferous period.

Lying southeast is the Blue Ridge Province. The deeply-eroded mountains of this region are composed of crystalline Pre-Cambrian and Cambrian rocks that have been strongly metamorphosed. They are, however, now separated from the Valley and Ridge rocks by a major dislocation – a thrust fault. This has carried gneisses northwestward from their original position and deposited them partly on top of the Palaeozoic succession.

Southeast from this lies the Piedmont Province, a region where Pre-Cambrian and Palaeozoic rocks outcrop. These originally were volcanic and sedimentary deposits which became deeply buried, then invaded by granitic magma. Subsequently they were highly deformed, then eroded, and finally thrust northwestward over the Blue Ridge rocks. Volcanism in this province commenced about 700 million years BP, and continued, more-or-less unabated, for about 200 million years. At least two episodes of deformation affected the rocks, one at the close of the Devonian, the other during the early Carboniferous period.

In contrast, the rocks of the coastal plain have remained relatively undisturbed. The oldest strata date from the Jurassic period (200 million years BP) and are underlain by rocks similar to those now found in the Piedmont region. The present continental shelf is a submarine extension of this marginal belt. Recent geophysical work reveals that the younger sedimentary rocks of the Valley and Ridge Province continue eastward and underlie the older metamorphosed strata of the Blue Ridge and Piedmont Provinces. The rocks of the latter two provinces evidently were metamorphosed before being thrust northwestward over the continental shelf sediments of ancestral North America, which seems to have grown by piling up thrust masses along its margins, at least during the lifetime of the Appalachian mobile belt.

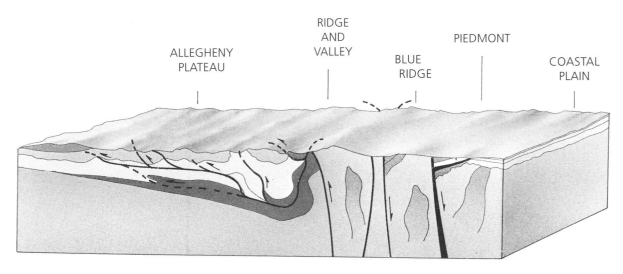

ALLEGHENY
PLATEAU

RIDGE
AND
VALLEY

BLUE
RIDGE

PIEDMONT

COASTAL
PLAIN

Present-day features of the Appalachian region. The vertical scale is, of course, grossly exaggerated.

Landsat image of the Appalachians in central Pennsylvania (false colour). This covers the entire width of the Valley and Ridge Province, a region characterized by alternating anticlines and synclines. To the west is the Allegheny Plateau, a region rising to between 300 and 900 m above sea level and built from gently-folded Palaeozoic sedimentary strata. Within the main fold belt the intensity of deformation increases towards the southeast (note how the valleys become increasingly narrow in this direction). Next comes the Great Valley, eroded out of Palaeozoic limestones, while to the east again (bottom right) is the Blue Ridge Anticlinorium, a resilient block of ancient Pre-Cambrian igneous and metamorphic rocks.

16

Late Palaeozoic mobile belts

Introduction

The late Pre-Cambrian and early Palaeozoic orogenic events ended in mid-Devonian times, by which time ancient North America and Europe were joined together and lay between latitudes 20° N and 30° S, spanning the Equator. The latter is believed to have passed through northern Norway, central Greenland, northern Canada and Alberta.

Currently, a broad zone of deformed rocks stretches across Europe and encompasses large parts of Spain, France and southern Britain. The same kind of deformed rocks extend through north Africa, via Iran and along the length of the Ural Mountains, before tracking along the north flank of the Angara Shield of Russia. Together these rocks were affected by the great Hercynian orogeny, which threw up new fold mountains across ancient Europe, Siberia and China. The late Palaeozoic Alleghenian orogeny which affected the southern Appalachians was an integral part of these events. As was the case with the Caledonian orogeny, new fold belts arose along the sites of earlier sedimentary basins which had formed along several continental margins.

The Hercynian mobile belt

Reconstruction of the palaeogeography of Europe at this time is rendered particularly difficult because of the later Alpine orogeny. This orogeny left its imprint on most of southern Europe and therefore superimposed its effects upon the regions bearing Hercynian fingerprints. Although there is much speculation regarding the plate movements which occurred at this time, the available evidence suggests that Europe and Africa were separated by an ocean during the late Palaeozoic, prior to the Variscan orogeny. As plates bearing Europe and Africa inexorably converged, this ocean was progressively destroyed, collision occurring during early Carboniferous times. The events which followed folded and metamorphosed the

The configuration of the continents during Devonian times.

deposits of the marginal basin and led to the intrusion of large granitic bodies into the underlying crust.

At this time, the southern margin of Europe was bordered by a series of large sedimentary basins; to the east similar basins lay adjacent to the ocean separating Europe from Siberia. This was to be the site of the future Ural Mountain chain. Similar basins existed along the western margin of the Pacific and spread down into what is now Indonesia. These collected large volumes of sediments, those found along the northern side of the Hercynian fold belt having been washed off the eroded Caledonian mountains of Europe. By late Devonian times, however, these had been lowered so much that they ceased to provide much debris, which instead came from the south where it is thought that several high plateaux existed.

The Devonian and Carboniferous marine deposits are predominantly shales and sandstones, with just a few limestones and volcanic rocks. This appears to have accumulated in a basin which shallowed towards the north, towards the continental platform.

Massive-bedded Old Red Sandstone strata, Wilderness Quarry, Forest of Dean, England.

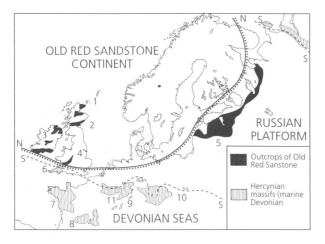

Map showing the outcrops of continental Old Red Sandstone (ORS) and marine Devonian rocks in western Europe: 1. Orkney; 2. Moray Firth; 3. Ireland; 4. south Wales; 5. Livornia Hercynian massifs of marine Devonian rocks; 6. southwest England; 7. Brittany; 8. Massif Central; 9. Black Forest and Vosges; 10. Bohemia and Silesia; 11. Ardennes. The line N–N indicates the furthest limit to which the Devonian sea advanced north over the ORS continent, while the line S–S shows how far south ORS interfingers with marine strata.

Corals flourished in the warm, shallow waters of the continental nearshore. Further north lay the continental interior, sometimes called the 'Old Red Sandstone Continent' on account of the *red beds* that collected on its surface under the tropical climatic conditions which prevailed at the time.

At the close of Devonian times similar red beds were being deposited in the Ardennes region of France and Belgium, which at that time was an island massif rising above the sea; similar island massifs occurred in parts of Germany and other parts of France. The sea then seems to have retreated from the region towards the close of the period, but read-vanced in early Carboniferous (Mississippian) times. It was during this latter stage that the extensive Carboniferous Limestone deposits of western Europe accumulated. These limestones are packed full of the remains of corals, brachiopods, crinoids and bryo-zoans.

While this was going on, Gondwanaland was slowly encroaching on Europe, a subduction zone

Weathering of Carboniferous Limestone at Buttertubs Pass, northern England. The soluble nature of this fossiliferous limestone leads to the formation of clints and grikes and of limestone 'pavements'.

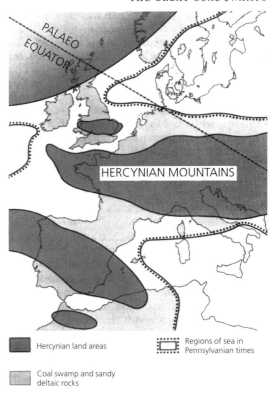

Map to show the Hercynian landmasses in Pennsylvanian times, together with the probable areas of marine deposition. Marginal coal swamps and sandy deltaic deposits are shown in grey. The palaeo-Equator ran through the northern tip of Scotland, the Low Countries and Hungary at this time.

existing between them. It is not entirely clear whether this dipped northward beneath Europe or southward under Gondwanaland, although the arrangement of the associated volcanic rocks tends to tip the balance in favour of the former arrangement. In this case an island arc lay off the southern margin of Eurasia during Devonian times.

The main episode of tectonic activity took place during the early part of the Carboniferous period, but it has to be noted that some parts of southern Germany and Bohemia were affected earlier, during the late Devonian. Once the principal orogeny got under way, the entire mass of sediments and volcanics were intensely folded, metamorphosed and invaded by granites. These granites have been dated as having been intruded in two phases: the first between 350 and 330 million years BP, the second between 290 and 280 million years BP. This orogenic episode terminated marine sedimentation and raised new fold mountains which extended right across the European region.

The great coal swamps 300×10^6 yrs

During late Carboniferous (Pennsylvanian) times, erosion of the European continent continued, huge volumes of land-derived sediment being flushed out into the marginal seas. Because the land had been lowered for a lengthy period by erosion, there was a great uniformity of geographical conditions over very wide areas, a fact which can be illustrated by noting that there is one particular marine horizon which can be traced from Ireland to Russia! As the land was lowered, so massive deltas built out from the land.

At the same time that this situation was developing, it has to be remembered that that much sea-water had been locked up in the polar ice of the late Palaeozoic ice age. As is typical of such periods of cold climate, glacial and interglacial periods alternated, with the result that there were continual changes in

sea level over relatively short periods. Such short-term changes may well have been responsible for the rhythmic style of sedimentation that set in in the deltaic regions. This may have been accentuated also by slight up-and-down movements of the land.

The typical rhythmic sequence which is found in the *cyclothems* of the British Upper Carboniferous, begins with a coal seam that formed from the remains of dead plants washed into the marginal deltas. Above the coal comes a shale horizon containing marine fossils; this indicates a deepening of sea level, with invasion of the low-lying marginal swamps by sea-water. Higher again in the sequence, shale with marine fossils passes upward into shale with a brackish- or fresh-water fauna: presumably lagoons had by this time developed. Higher still the shale gives way to sandier deposits, these eventually showing a transition into sandstone with many plant remains. The change to this coarser material is taken to imply either that the land behind the deltas had beome somewhat higher, leading to enhanced erosion and production of coarser sediment, or that river-borne sediment was being dumped over the site of the former lagoons due to the increased width of tidal flats.

Legend (map key):
- Hercynian land areas
- Coal swamp and sandy deltaic rocks
- Regions of sea in Pennsylvanian times

Reconstruction of a Carboniferous coal forest. This would have been dominated by seedless trees and seed ferns. [Photo: British Museum]

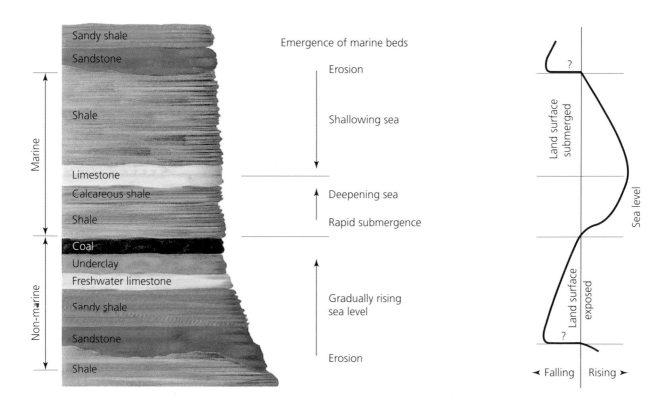

Section through a typical cyclothemic unit from the Coal measures. On the right is a graph relating the different beds to changing sea levels.

At the top of the cyclothemic unit the sandy beds pass into a coal seam, immediately beneath which usually is what is called a seat earth; this represents a fossilized soil horizon in which the roots of the swamp plants grew.

Rhythmic sequences of this kind are repeated many times over within the Upper Carboniferous succession. While the coal horizons represent but a relatively small part of such sequences, they are sufficiently important to have propped up the economy of many countries for decades. The gradual reversion to coal formation at the end of each cycle indicates that either the impetus for uplift (or drop in sea level) gradually died away, the swamps once more encroaching over the tidal flats and lagoons.

Events elsewhere

The main phase of uplift and deformation in the Uralide belt was later than in the rest of Europe, i.e. between the Late Carboniferous to Mid-Permian. Faulting in this region continued right through until the Triassic period. Meanwhile, in the region of the future Atlantic ocean, rifting allowed huge volumes of Permian basalts to rise to the surface; these are now exposed around Oslo in Norway and in the Midland Valley region of Scotland.

During the Permo-Triassic period most of Europe and northwest Africa was a land area, with non-marine red beds being laid down over wide areas. Their presence in the succession indicates that the climate was warm and sometimes arid. This situation found a parallel in North America at the same time. Finally, in northern Europe two brief incursions of the sea led to accumulation of salt deposits (evaporites), formed as the sea evaporated in more arid phases. Throughout the period there were many similarities between North America and Europe; for instance, the plant remains and reptilian fossils are remarkably alike, while red beds of similar age occur on both continents. Carboniferous coal swamps also are common to both, while the main episode of deformation occurred at much the same time too. Together this provides evidence that there was a continuous mobile belt prior to Triassic times.

Recent field work has shown there to have been a mobile belt of much the same age in West Africa. The succession within this Permo-Carboniferous belt contained large thrust slices which were moved eastward during orogeny. In many ways there is a parallel here with what was happening on the northern flanks of the Appalachians at this time.

The ancient Siberian shields also were bordered by sedimentary basins. While the Uralide mobile belt ran along the western edge of Siberia, it was not the only one: thick sequences of marine sedimentary and volcanic rocks are found in a zone extending south of the Angara Shield and running west–east through the southern tip of Lake Baikal. Red beds typify the northern part of the belt, these ranging in age between Devonian and Early Carboniferous; southward they pass into limestones and marine shales.

All of these strata were folded during early Carboniferous times, further deformation taking place during the Permian period. Similar mobile belts also surrounded ancestral China, these being aligned roughly along the trend of early Palaeozoic basins. One of the latter – the Central Asia basin – was raised up into a mountain chain during the late Palaeozoic, presumably as ancient China collided with Siberia.

Events in the interior of Siberia were not dissimilar to those prevailing in Britain at the same time. In the west, Devonian red beds accumulated, passing eastwards into marine limestones. The later Carboniferous and early Permian strata are mainly limestones but there was also a significant accumulation of coal. The coal contains a flora that is rather like the *Glossopteris Flora* of Gondwanaland (see Chapter 17), but palaeomagnetic data indicate these coals were formed at somewhat lower palaeolatitudes. Later in Permian times there was widespread volcanism over the western part of the continent. The story of China was rather similar.

Mineralization and plate tectonics

By the time the Variscan orogeny had ended most of Europe's mineral wealth had been created. In recent years it has been realized that the creation of this source of wealth has close ties with plate tectonics and orogenies. Although some metals are concentrated within stationary bodies of magma – simply by the settling of dense metal-bearing crystals (e.g. chromium) – most ore deposits are concentrated rather differently. For instance, copper, tin and molybdenum crystallize from solutions associated with granitic magmas.

Most sulphide ores (and many kinds of ores are found in this form) appear to have been precipitated from hydrothermal solutions: hot fluids that have the capacity to carry metals in solution. Such fluids origi-

nate in magmas of various kinds and then rise through the overlying crustal rocks, carrying with them dissolved metals from the magma source. The dissolved metals are precipitated after complex chemical reactions between the wall rocks of the cavities through which the solutions pass, and the solutions themselves. The solutions may also attack the rocks through which they pass, leaching out metallic elements and redepositing them elsewhere. The metal-charged solutions are called mineralizers.

Some of the world's richest sulphide deposits are located under the Red Sea. This formed along the line of an ancient rift which was reactivated by plate divergence during the Tertiary period. It is a part of the rifting that has opened the Red Sea, Gulf of Aden and the East African Rift Valley. The ores are found at depths of 2 km below sea level, in sediments which are between 20 and 100 m in thickness. The concentrated metals include iron, zinc, copper, lead, silver and gold. The pore spaces between the mineral grains in the sediments are saturated with brines which carry the same metals in solution. Presumably these

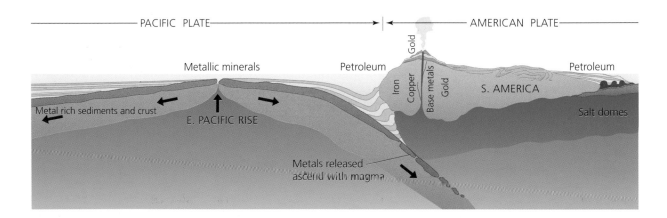

The relationship between plate boundaries and mineralization. The Pacific plate is being subducted beneath the western margin of the American continent. Sediment and crustal material enriched in metallic ores by activity along the East Pacific Rise are partially melted. Metals rise with the magmas produced, creating the metal-bearing rocks of the Andes.

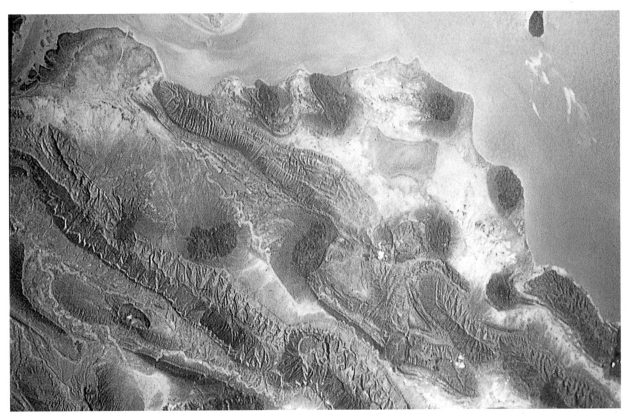

Anticlines and salt domes near Bandar Abbas, Iran. These folded rocks form a part of the Zagros Mountains, and are a valuable source of hydrocarbons. [Gemini photo: courtesy of NASA]

metal-charged brines circulated through the porous sediments above the spreading axis – an obvious site of magmatic activity and high heat flow.

The Red Sea is a relatively young ocean – indeed, one of the youngest on Earth; it marks the suture where a new wave of ocean creation is taking place. What, we may reasonably ask, is the situation in the vicinity of the spreading axes of well-established oceans?

The rocks of oceanic rises are intensely altered by heated sea-water in nearly every site that has been satisfactorily researched. It seems that hydrothermal solutions, rising from depth in the mantle, penetrate fissures in the oceanic crust, dissolving metals out of the sea-floor rocks and finally concentrating them as ore deposits. Sea-floor sampling shows that iron, manganese, copper, cobalt, chromium, uranium and mercury are especially enriched in such localities.

One particularly valuable ore concentration is found among the Troodos Mountains of Cyprus, where ancient ophiolites have been thrust onto the old continental margin, then buried by more recent strata. Study of ophiolites allows us to investigate, very conveniently above sea level, a slab of oceanic crust that once originated at a spreading axis. The valuable copper, iron and chromium ores of this region are closely related to the volcanic rocks in which they occur. Thus, interbedded with the volcanics are manganese-rich and iron-rich horizons that are identical to the mineralized sediments dredged from the ocean floor at mid-oceanic ridges.

Metalliferous deposits of different kinds are concentrated along convergent plate boundaries, where oceanic crust is being subducted beneath either continental crust or an island arc. The processes that operate under these conditions are complex and very diverse. The valuable ores of Japan, the Phillipines and the Cordilleras of the western Americas, are located along such modern boundaries, while the older ore deposits of Zimbabwe, West Australia and Alaska are found along the lines of old convergence zones. The simplest way to explain what happens in these situations is to assume that mineralizers arise from the subducted slab as it plunges towards the hot

(a)

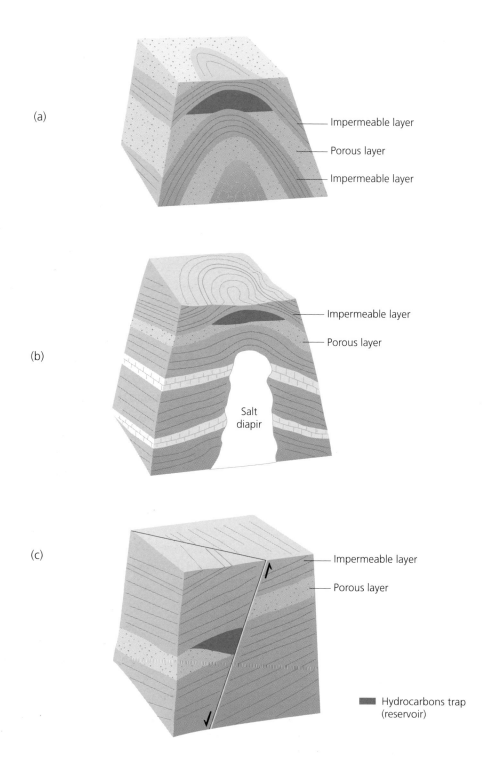

Impermeable layer

Porous layer

Impermeable layer

(b)

Impermeable layer

Porous layer

Salt
diapir

(c)

Impermeable layer

Porous layer

Hydrocarbons trap
(reservoir)

Various kinds of oil trap: (a) anticline; (b) salt dome; (c) fault.

mantle. Because this may involve continental as well as oceanic crust, the chemical conditions have the capacity to be more diverse in these locations.

Hydrocarbons

Here is as good a place as any to investigate the formation of hydrocarbons, for this is also related to lithospheric plate activity. Rifting and plate convergence have been shown to be capable of providing most of the key factors for the generation and accumulation of oil and gas.

Crude oil, or petroleum, is the most valuable and versatile of Earth's buried treasures. It provides us with a number of different fuels, each of which is essential to modern civilization. More recently, natural gas has also emerged as an important resource, and in some parts of the world is piped directly into people's houses and factories. The first question, therefore, is exactly how are oil and gas formed?

Oil was formed from the remains of algae and plankton that lived in the oceans millions of years ago. Gas is formed in much the same way. At some periods in Earth's history, productivity of the organisms was higher than normal and they accumulated on the sea bed and became buried under layer upon layer of marine sediment. Where burial was rapid and the remains did not have time to be oxidized on the sea bed, conditions were ripe for oil formation. Subsequently the remains were broken down by bacteria and converted into a substance called kerogen. Over a lengthy period of time, further layers of sediment buried the organic material deeper and deeper, causing both the pressure and temperature to rise. This cooked the kerogen until it generated many different kinds of hydrocarbon.

Once the fluid hydrocarbons had been formed in this way, they occupied pore spaces within their host rocks. However, being relatively light and volatile, they tended to rise towards the surface of the Earth via capillary action. Finally, upon reaching an impermeable horizon, they ceased to rise and become trapped, forming a reservoir of oil and gas. Normally this floats on a 'pore-sea' of water, hydrocarbons being less dense than the normal pore-water in sedimentary deposits.

The trapping of the oil is a vital step; without it the hydrocarbons would simply rise to the surface and escape. Deformation of the overlying rocks is a vital step in the entrapment process, anticlinal fold crests being favoured reservoir cappings, as long as some of the anticlinal crests are composed of impermeable strata. Salt domes can also cause traps, since as they rise upwards they deform the adjacent strata, providing suitable traps in the process.

It is now easy to see why plate tectonics plays a part here. First of all, suitable shallow seas and shelf sediments are required, so that algae and plankton can thrive, then die and settle into the marine mud below. These conditions would be typical of a continental margin or a shallow shelf sea. Secondly the ocean needs to be closing, so that eventually the oil-bearing sediments and their reservoir rocks become rucked up, forming suitable traps. What better place than a closing ocean? Such a situation applies to the Persian Gulf, where huge reservoirs of oil are found among folds in the Zagros Mountains. These were thrown up as the ancient continental block of Arabia was driven against Eurasia as part of the break up of ancient Gondwanaland.

The Permian mass extinctions

Of all the periods in Earth's history, the end of the Permian saw the greatest change in the number and balance of animal families, genera and species ever seen. Beginning about 255 million years ago, roughly 95 per cent of all marine species vanished within 10 million years. Corals, brachiopods and ammonoids suffered the most, being especially severely depleted in tropical waters. The rock record suggests that extinction came in waves, rather than at any one point in time. Why did this occur?

We have to remember that in Late Permian times there was a single supercontinent which gradually emerged as plate movements took place. If we go back prior to 230 million years BP, shallow seas accounted for about 43 per cent of the total continental area; these shrank to 13 per cent by that time. This sharp reduction in sea area was parallelled by a decline in the diversity of species living. However, this alone would not have been enough to explain the very dramatic wiping out of life recognized from the fossil record.

Another suggestion that has been made is that the supercontinent, having been entirely surrounded by a major ocean, provided too small a number of distinctive shallow-water marine environments to support the diversity of life that currently exists (calculation suggests there would have been less than one half the modern number). If this were so, then there would be a decline in species diversity and numbers. A further proposal seeks to explain the extinctions by high-latitude cooling of the Earth – remember there was a Permian glaciation. However, the Permian ice age in reality was a rather minor affair. It has to be said, therefore, that currently we are nowhere near finding a satisfactory explanation for this major event. (This is one of the things which makes geology such an exciting science!)

17

Gondwanaland evolves

Gondwanaland in the Early Palaeozoic

During the early part of the Palaeozoic era, present day South America, Africa, Peninsular India, Antarctica and Australia formed a single supercontinent – Gondwanaland. This occupied a rather different position from the other continents, palaeomagnetic data indicating it to have occupied a southerly position. The ensuing chapter traces the evolution of this supercontinent, not only in Palaeozoic but also in more recent times. The break-up of this huge landmass left widespread evidence amongst the rocks of the modern continents and had a profound influence on Mesozoic and Cenozoic events. The painstaking piecing together of the various shreds of evidence has enabled geologists to write the story that follows.

During the earlier part of the Palaeozoic, Gondwanaland appears to have been virtually surrounded by subsiding basins in which were deposited marine sedimentary rocks now found in such widely

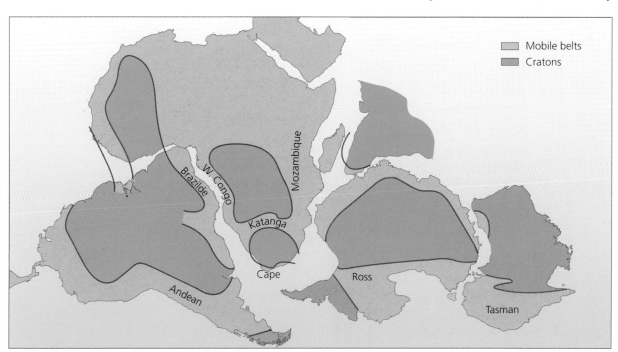

Gondwanaland in the Early Palaeozoic, showing mobile belts in grey and cratons in yellow.

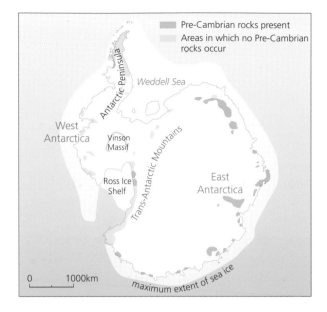

The general geology of Antarctica.

separated places as east Australia, New Zealand, north India, north and west Africa, Florida, the Andes, South Africa and Antarctica. Large remnants of ancient cratonic cores – once the heart of Gondwanaland – are to be seen in Africa, Central America, Brazil, parts of the southern USA, Pakistan and India. Palaeomagnetic and palaeoclimatological evidence from the rocks of the old continental interior shows the supercontinent to have stretched across a wide range of latitudes during this period.

The continental interior

Among the sediments that accumulated on the cratonic regions are distinctive sandy horizons which contain both facetted and striated pebbles and boulders. Both features are typical of glaciated materials, i.e. tillites. The tillites are of Ordovician age and have been discovered in the Saharan region of north Africa, and also in South America. It is probable that the boulders were dropped from drifting ice, possibly in the form of vast continental ice sheets, which spread across this region at that time. Such an interpretation is supported by geophysical data which show that during the Ordovician the South Pole lay in west central Africa. All of these glacial rocks were apparently located within 50° of the palaeopole.

The central part of Australia was at this time a major depositional basin, and there were smaller

ones on the northern and western perimeters of this continent. The rock succession found in these basin assemblages includes sandstones, shales, limestones and evaporites. Similar evaporites are found in northern Pakistan where, in the Salt Mountains, are 400 m of salt and gypsum deposits. Clearly both India and Australia experienced an arid climate during at least a part of the Lower Palaeozoic. This conclusion is supported by palaeomagnetic information, and contrasts sharply with the glacial conditions being experienced by Brazil and Africa at the time.

The southeastern part of the USA and southern Mexico were both a part of Gondwanaland during this time. Wells drilled in southeast Alabama, southern Georgia and northern Florida have revealed the existence of Lower Palaeozoic volcanic and sedimentary rocks in which the fossils are similar to those found at the same horizon in northern Africa.

Margins of the supercontinent

At the commencement of Palaeozoic time Gondwanaland was girdled by subsiding basins that eventually evolved into orogenic zones. The shallow, nearshore waters were collecting grounds for clastic sediments and shelly limestones, and the living environment for abundant calcareous oganisms. The deeper, offshore waters were characterized by black shales and turbidites, together with volcanic rocks derived from island arcs; the fauna was dominated by planktonic graptolites. The remnants of this succession of rocks are now found in widely separated locations, and provide us with vital evidence in favour of continental drift.

Part of this circum-Gondwanan orogenic system now occupies the Andean chain. In early Palaeozoic times the more westerly of these South American basins collected thick marine sediments and volcanic rocks, the latter not being found further east. In places the sequences are 3000 m thick. Glacial rocks of early Ordovician–mid-Silurian age have been reported from both western Bolivia and southwest Argentina. Towards the close of the early Palaeozoic the peripheral basins were squeezed and deformed by plate collisions, probably by the subduction of oceanic lithosphere beneath an island arc or arcs situated some way off the continental margin, with eventual collapse of those arcs.

Another segment of the Gondwanan belt runs through the Himalayas, Pakistan, Iran, Turkey, Greece, Yugoslavia (or what was), Italy and north Africa. Between Cambrian and Devonian times this series of basins was joined with those that bordered the western flank of Africa. Here few volcanic rocks are to be found, but clastic sediments are; they may reach a thickness of 5 km. This tract was not affected by deformation during the Palaeozoic. Evidently plate convergence had not affected this margin of the supercontinent at this stage.

Finally, Lower Palaeozoic marine sedimentary rocks also are found along the coastal regions of west Africa and in the southern Appalachians. In the latter region, a 10 km thickness of shales and volcanic rocks is found in a belt extending from Carolina through Virginia to Georgia. The same rocks doubtless occur elsewhere on the craton but are covered by younger strata. Volcanism also occurred during the time of sediment accumulation; the rocks are typical island arc volcanics. It is assumed that a subduction zone plunged southeastward beneath the supercontinent at this time.

The Late Palaeozoic ice age – 1

This was a time of profound change; not only were there distinctive climatic trends but also the continents were coverging until, about 280 million years ago, one great supercontinent – Pangaea – existed. Prior to that, Gondwanaland had been the largest land area on Earth, extending from the Equator to the South Pole. Laurasia, situated in northern latitudes, was then in two parts, these being separated by the Tethys Ocean which widened to the east. It included at least two large islands. the westerly one was Laurentia, the more easterly is often called Angaraland.

In Chapter 12 we saw how, on the North Amercian craton, there were repeated fluctuations in the relative levels of land and sea, particularly during late Carboniferous (Pennsylvanian) times. A similar situation appears to have prevailed on Gondwanaland too. During this period there were repeated incursions and regressions of the sea, reflected in cyclothemic deposits on the borders of the ancient continent. Very reasonably, it has been postulated that these fluctuations occurred in response to changes induced by pulsatory ice-sheet growth. The ice is believed to have been centred on the late Palaeozoic South Pole. At that time this lay within Gondwanaland. We know that very rapid sea level changes occurred during the last (?present) Ice Age, therefore there is every reason to suspect that such would have happened in the distant past too.

What was the world like during this period? Well, it certainly was quite different from today; for a start, the land surface was much less luxuriously clothed in vegetation, although a rich flora existed beneath the water of the oceans. There is a strong possibility that climatic zones were established, much in the same way as they are today, although overall temperatures may have been slightly lower. Then, in Pennsylvanian times, as the various cratons converged, the situation changed. By the middle of Permian times the sea separating the components of Laurasia had been eliminated, the new supercontinent gradually rotating until colliding with Gondwanaland, fusing the two together into one vast landmass: Pangaea.

The collisions involved in this coming together created new chains of fold mountains along the sites of old sedimentary basins. These lofty barriers, together

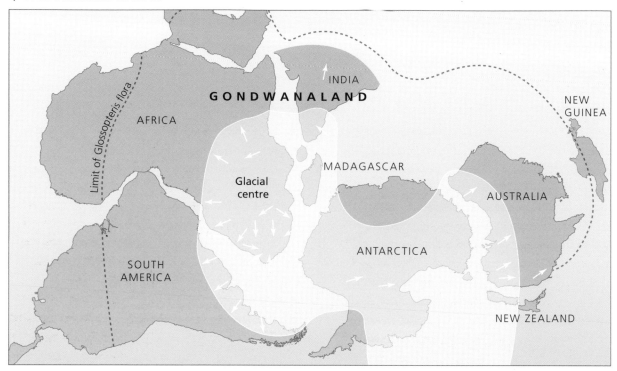

The Late Palaeozoic ice-sheets of Gondwanaland. The centre of glaciation shifted during Permo-Carboniferous times; thus southern Africa and South America were affected first, with eastern Australia and Antarctica affected last.

with the global geographical changes, had a profound effect on ocean currents and atmospheric circulation, changing the Earth's climate as a result.

Extensive volcanism accompanied the tectonic activity and great sheets of lava spread out along the continental margins. There is evidence that snow-fields capped the higher mountains, not only in middle latitudes but also in the equatorial zone. Global cooling was the ultimate result, and extensive ice-sheets grew over much of Gondwanaland. Glacial deposits and the fossilized imprint of glacier activity can be found on all of Gondwanaland's continental pieces. They provide us with one of the most powerful pieces of evidence for continental drift.

The Late Palaeozoic ice age – 2

There is little doubt that there were alpine glaciers in both Bolivia and Argentina, as early as Devonian times; subsequently they became far more extensive and there were at least two major glacial episodes during the Carboniferous period. In Brazil, roughly 1500 m of glacial beds accumulated between 320 and 270 million years BP. Detailed studies of these beds show that there were at least 17 glacial–interglacial cycles during the 50-million-year period. When fea-

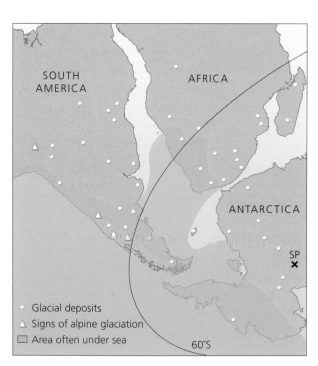

- • Glacial deposits
- △ Signs of alpine glaciation
- ▢ Area often under sea

SP
×

60°S

Map to show the possible site of the maritime part of the great ice-sheet. Over most of this region the base of the ice-sheet was close to or beneath sea level, rendering it rather unstable.

tures produced by the moving ice are mapped – for instance glaciated rock pavements and striated valley walls, they indicate that the main ice-sheet lay off the eastern coast of South America. The direction of ice motion was from east to west. This may seem odd at first, but when we take into consideration the configuration of the continents, it becomes more acceptable.

The African and South American cratons were adjacent at this time. In southwestern Africa there are significant glacial deposits of Permo-Carboniferous age. The most famous is the Dwyka Tillite, which runs across southern Africa into Madagascar. The glacial strata are over 1000 m thick in the Transvaal which appears to have been one of the principal areas of ice dispersal. In the highest part of the succession are plant-rich beds that contain a flora dominated by the seed-fern, *Glossopteris*. It also contains remains of the aquatic reptile, Mesosaurus, a Permian creature found only here and in Brazil.

Glacial rocks of the same age are also found in Australia, where the earliest signs of climatic change were felt in the mountains of Tasmania and southeastern Australia. Subsequently the ice-sheet spread as far as northern Queensland. The wide distribution of tillites indicates that most of southern Australia must have been covered by ice at least once during this period. Features left behind by the retreating ice indicate that the main movement of ice was from the south. At this time apparently Antarctica occupied this position, a notion which posed severe problems for geologists not many years ago. The reason was that no glacial deposits of comparable age had then been found on that continent; however, in 1960, extensive glacial beds were discovered in the Transantarctic Mountains and elsewhere. In places these are almost a kilometre in thickness. Such a massive occurrence emphasizes how Antarctica was a major source of ice during Permo-Carboniferous time.

India and Pakistan must also be included in this part of the story. The northern margin of India was at this time a part of the southern coastline of the Tethys Ocean. Upland regions further south were apparently the source for extensive ice-sheets, which sometimes spread into the sea. The glaciation here spanned the interval 310 – 270 million years BP, and there were at least three glacial episodes. Glacial rocks also are found in the Falkland Islands.

Of necessity, the details of the glaciations have been much simplified. It is, however, widely accepted that southern Africa was glaciated for the longest period. It is also accepted that, as Gondwanaland gradually shifted its position with respect to the South Pole, so the focus of glaciation shifted. Together all of the evidence indicates that southwest Africa and South America became glaciated first, followed by South Africa, Madagascar, India, finally moving on to Antarctica and Australia.

The rise of plants

Putting all the evidence together, it seems that the Late Palaeozoic was the period when plants really became abundant on the land surface of the world. We know that the climate of the Late Carboniferous tropics was sufficiently warm for the development of luxuriant coal swamps at the continental margins. Inland, however, conditions appear to have been arid. Then, in Early Permian times it is possible that ice-sheets pushed to within 30° of the Equator and it is likely that there had to be a compression of the Earth's climatic zones. Any life endeavouring to survive in regions marginal to the great ice-sheets would have had a particularly tough time, and only the more adaptable species would have survived the rigours of glaciation.

Enormous reserves of coal bear witness to the huge quantities of dead plant life that lived and died on

Glossopteris (Gondwana realm).

Lonchopteris rugosa (Amerosinian realm).

Nothorhacopteris argentinica (Gondwana realm).

Angaridium finale and *Gondwanidium sibiricum* (Angara realm).

the continental margins. The Carboniferous coals of North America, Europe and Asia grew under tropical conditions; however, the coals of the southern continents survived despite much cooler conditions. The principal contributors were land plants.

Study of ancient plants shows that primitive vascular plants had evolved by the Late Silurian period (about 400 million years BP); by Late Devonian times these had evolved to form extensive lowland forests of tree ferns with a 10-m-high canopy. They spread quite rapidly over the northern hemisphere. By the commencement of Carboniferous times they were the chief contributors to the peat and coal swamps. Fossilized remains suggest the ferns attained heights of 20 m in some cases.

The earlier, more primitve, plants reproduced by

means of spores, but the seed-ferns which flourished during the Carboniferous period were the first to regenerate via seeds. These were to become the stock from which the later cycads, conifers and flowering plants evolved during the Mesozoic era. The most important coal forest plants were the Lycopsids and Sphenopsids which stood tall in the swamps of the time. The former are today represented by the low-growing temperate forest club-mosses, while the latter's only descendant appears to be the curious little horsetail, *Equisetum*.

The various continental floras were quite diverse, and sufficient is known about them to recognize different plant 'realms'. Thus the floras found in the region of the Angaran Shield of ancient Siberia are preserved in coals that were deposited at around latitude 45° N. Palaeomagnetic data show that although similar plants were found in Europe and North America, they were flourishing at around palaeolatitude

10° N/S. The Glossopteris Flora of Gondwanaland, on the other hand, lived in latitudes higher than 30° S. Thus three plant realms are recognized: the Amerosinian (which includes floras of the equatorial belt), the northern Angaran, and the southern Gondwanan realm. There are significant differences between them, the restriction of *Glossopteris* to Gondwanaland being another piece of the evidence supporting the drift and break-up of the super-continent.

Late Palaeozoic animals

The animals of the early part of Palaeozoic time were rather different from those of the later. For instance,

after the Devonian period, trilobites became very rare and were extinct by the Late Permian. Snails (gastropods) do not appear to have been very abundant but the coiled ammonites started to increase in numbers and were widely distributed. These ammonites pro-

Freshwater lamellibranchs (Coal Measures).

Pentacrinites, a Jurassic crinoid.

Dapedius, a Devonian fish.

Brachiopods, bryozoans and their spines on a small slab (Upper Permian).

Infilled plant stem (Pennsylvanian).

vide geologists with vital 'zone fossils' for this period of geological time. The brachiopods continued to evolve, spiny-shelled forms appearing. The devlopment of spines assisted in anchoring them to the sea floor and helped in straining food from the mud-laden water; it also afforded them some protection from their enemies.

During the Devonian period reef-forming groups like the stromatoporids, corals and bryozoans became important. This changed during the Carboniferous, when crinoids (sea-lilies) took over the lion's share of reef-building activities. They became not only abundant but also very diverse, and have remained so until the present day. Of the smaller organisms, the protozoans underwent tremendous evolutionary changes. Foraminifera diversified and proliferated during this period. Many of the genera evolved very rapidly and were apparently highly specialized; they provide palaeontologists with ideal zone fossils.

One of the most significant features of Late Palaeozoic life was the colonization of the land. In the Late Devonian the first amphibians appeared; these were fish-like but very quickly developed so that their bodies became flattened and their eyes moved towards the top of their head. Presumably they lived in shallow water. Some adapted to land by redesign-

ing their backbones and legs, but they tended to be rather clumsy and, in any case, had to return to the water to lay their eggs. In contrast, the reptiles, whose eggs have a tough skin, were able to survive more readily away from the water.

The earliest reptiles looked little different from the amphibians, but during Carboniferous times they evolved quickly and some very bizarre forms inhabited the low-lying swampy plains. One group – the Pelycosaurs – had long spines with skin stretched between them. These probably were used to regulate their body temperature in some way. They became so specialized that they apparently could inhabit only a narrow ecological niche, and were extinct by the close of Permian times. Other groups were much more successful and some of the mammal-like genera gave rise to true mammals that flourished during the Mesozoic era.

During Late Carboniferous times there was a rapid increase in the numbers of insects. Abundant fossils of large cockroach-like insects and enormous dragon-flies have been recovered from lacustrine and lagoonal deposits within some of the coal-bearing sequences. Hitherto insects had existed, having made their first appearance in Silurian times, but it was only now that they evolved and proliferated rapidly.

18

Stirrings in the Alps

Introduction

The Mesozoic era saw the final break-up of Gondwanaland, together with the development of the Alpine orogenic cycle. The beginnings of the latter were seen in the growth of major troughs on or closely alongside the older Hercynian belts, and ended – at least in the region of Europe – with the uprise of the Alps and Carpathians during Tertiary times. Orogeny also affected Asia; the Earth's greatest mountain chain, the Himalayas, being thrust up a little later, but during the same general period of orogenic activity. Movement in this region is still continuing.

The Alpine cycle takes us from the events of the Late Palaeozoic up until the present time. The huge upheavals the Earth's crust experienced gave rise to a very complex sequence of events which accompanied the approach of Eurasia and Africa, with subduction

Typical view of the European Alps: the Jungfrau, Monck and Eiger from near Interlaken, Switzerland.

Section through the western part of the Alpine chain.

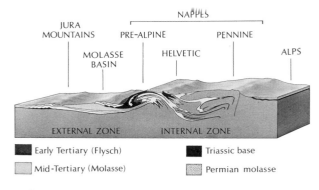

JURA
MOUNTAINS

NAPPES

PRE-ALPINE   PENNINE

MOLASSE
BASIN

HELVETIC

ALPS

EXTERNAL ZONE   INTERNAL ZONE

■ Early Tertiary (Flysch)   ■ Triassic base
□ Mid-Tertiary (Molasse)   ▨ Permian molasse

of the intervening oceanic crust. It was during this period that the great oil wealth of the Middle East was laid down, the Red Sea and Gulf of Aden opened, and the Mediterranean went through contortions which began its gradual but inexorable closure.

The Alpine cycle

The Alpine cycle really began about 240 million years ago, in the Triassic period. Marine shales and carbonate sediments were laid down in basins which had developed among the eroded stumps of the ancient Hercynian mountains. Triassic and Permian sedimentary rocks are located in a broad belt that includes the Alps and Carpathians – one of the most closely-studied regions of the world. It was among these mountains that European geologists first began to perceive the exact way in which fold mountains developed.

In addition to the marine sediments, carbonate reefs grew in the regions of Austria and northeastern Italy. Things were very different in northern parts; for instance in Germany and Poland not reefs but continental deposits were being formed. This is taken to indicate that the present Alpine region lay near to the southern margin of Mesozoic Europe, while island arcs built across the region, the reefs being associated with these, much in the same way as they are in modern Indonesia, for instance.

During the Jurassic period (commenced 180 million years BP), the sea expanded northward, flooding much of Europe. It reached a maximum extent during the Late Cretaceous, when it covered the whole of Europe and Russia, extending also across North America, North Africa and Arabia. Without doubt this was one of the most extensive marine episodes of all time; because it occurred during a particular subdivision of the Cretaceous period, it is termed the Cenomanian transgression.

By mid-Cretaceous times the crust in the Alpine

region had become rather unstable and a number of elongated basins developed; these were punctuated by large islands. Sporadic crustal uplift alternated with subsidence and periods of rapid sedimentation, presumably as short pulses of tectonic activity affected the region. A similar style of geological activity has now been recognized to have taken place in other orogenic cycles; it is called *flysch* sedimentation. Flysch basins developed both within the folded zone and at its margin, immature turbidite-type sediments accumulating rapidly.

Eventually, during the early part of the Cenozoic era, the geography of these large islands became more and more different as the shallow seas began shrinking; Europe and Africa were approaching one another. During the Oligocene – Miocene period collision occurred, the mobile zone shuddering under the culminating Alpine orogeny. This threw up the buried sediments into great folds, raising up a new fold-mountain chain whose remains we can still clamber over today.

The main orogenic phase strongly deformed the sedimentary deposits, not only the Mesozoic ones but

Crumpled flysch beds, Sicily.

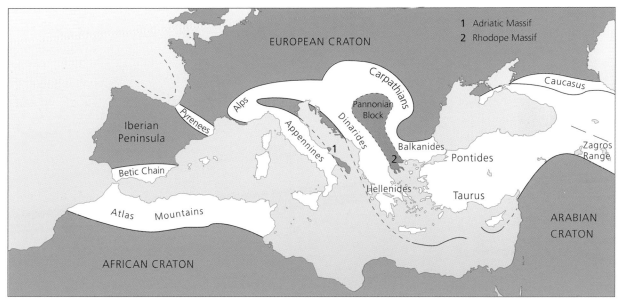

Generalized map of the Alpine chain and its eastward extension into the Caucasus and Zagros Mountains of Iran. The complex structure of the Mediterranean region indicates that substantial rotation of parts of southern Europe must have occurred during Tertiary times, since the main fold belt was produced (1. Adriatic Massif; 2. Rhodope Massif).

also the underlying Hercynian rocks. Magma bodies rose into the crust, giving rise to large granitic plutons and masses of more basic magma. The uplift of what is now Switzerland caused immense slabs of only partially coherent rocks to slide over each other, and pile up. Where deformation affected somewhat more coherent strata the folding was of a more plastic kind, the rocks being rucked up into great folds which often flopped over to become flat-lying, or *recumbent*. These flat-lying structures, first described

Tectonic map of the Alpine mobile belts extending into the Middle East and central Asia.

from the Alps, became known as *nappes*; in many instances the individual fold limbs were torn from their roots and transported tens of kilometres from their original positions.

While this was going on there was regional elevation of the European craton, the sea being expelled from the central Alpine tract. The very vigorous erosion that followed the rise of the new ranges, saw rapid deposition of debris in non-marine basins adjacent to the tectonized zone. This later phase has become known by the term *molasse*.

Metamorphism accompanied the deformation of the central regions. Radiometric dating shows that the majority of the Alpine nappes had been emplaced by around 70 million years ago; the metamorphism followed shortly afterwards. Both were a function of the elimination of oceanic crust as the Eurasian and African cratons converged. Since the orogeny finished the Alpine chains have been deeply eroded. By Pliocene times much of the region was probably of quite low relief. The magnificent mountain scenery of today is largely the result of uplift and erosion just before the Pleistocene epoch.

Flysch deposition

During the early stages of the cycle, sedimentation

Flysch outcrop, near Cefalu, Sicily.

(a)

(b)

(c)

(a) Plan view of flute casts in greywacke beds. The teardrop-shaped features are the casts of vortex marks produced by swirling mud-laden currents in the local sea-floor. The orientation of the casts indicates the general direction of the palaeocurrent flow, the deeper ends of the casts showing which direction the turbidite flowed from. (b) Elevation view of the above features. (c) Groove casts in turbidite bed. These striations were produced by rapidly-flowing currents which etched grooves into the muddy sea-floor by the abrasive action of sand grains and pebbles.

along the continental margins tended to be dominated by chemically-precipitated rocks, e.g. limestones. Later, as the conditions became less and less stable, the nature of the sediments changed dramatically and land-derived debris became the norm, this either becoming mixed with the carbonates, or entirely replacing them. The facies change occurred in Cretaceous times, the resultant deposits being of Late Cretaceous and earliest Tertiary (Palaeocene) age. Their deposition signalled the commencement of the orogeny proper in the core regions of the belt.

The typical kind of deposit is termed *flysch* (from a German dialect word) and the thickness of flysch in the region is seldom more than 3 km; however it is very widespread. In different places the flysch is of different age: we say it is diachronous. The earliest flysch was generated where the earliest orogenic disturbances were felt, i.e. in the internal zones close to the orogenic core. The younger flysch sequences tend to be located in the external zones, i.e. those regions furthest from the orogenic core, where they may be of Palaeocene or even Eocene age.

The main source of flysch was the rising orogen. Thus, as folds developed and the crust thickened as a result, the more elevated regions were rapidly attacked by erosion and the detritus strewn out quickly by streams and rivers. Very rapid deposition meant that a complete jumble of different rocks types, coarse and fine, was deposited side-by-side. Earthquakes would constantly have destabilized the region with the result that the unstable sediment piles often slid off basin slopes as turbidity currents, being spread out on the deeper sea-floor as turbidite sandstones.

Certain features found associated with such rocks can be used to trace in which directions the palaeocurrents flowed, giving some idea of the source direction. These 'markers' suggest a very complex pattern of deposition, as short-lived ridges within or on the margins of the various basins, contributed not only

sediment but also constrained current flow in a very complicated regime. In both the Carpathians and the Hellenic ranges, redeposition of the flysch debris was largely accomplished by the activity of turbidity currents, which flowed axially along the major basins.

Molasse

The upheavals of Cretaceous times largely terminated the production of flysch, since they had the effect of eliminating the surviving areas of open sea, except in parts of the Aegean. By Oligocene or Early Miocene times, large tracts of upland has risen, much by the process of block faulting. These young massifs were soon attacked by erosion and the debris produced

Molasse-type pebble beds and sandstones.

Map showing the Carboniferous and Permian seas that invaded the European craton. The main marine area covered much of Russia and extended through Greece, Spain and North Africa. An arm also extended into Germany and Britain. This has become known as the Zechstein Sea and in it distinctive dolomitic limestones of Permian age were deposited.

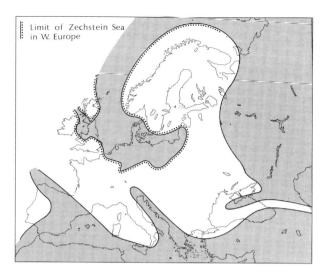

Limit of Zechstein Sea in W. Europe

was transported either into low-lying basins on the margins of the old orogenic belt, or among the new block massifs. Some of the old basins, now of course filled with sediment, underlie the European lowlands and marine gulfs of the present day. The sediments which accumulated in this post-orogenic stage are called *molasse* deposits.

Molasse is usually unconformable upon older strata which have been folded and metamorphosed; thus it clearly post-dates the climactic orogenic phase. Much of the sedimentation shows a rhythmic pattern rather like that of the Coal Measures. Typical molasse includes non-marine conglomerates, sandstones and shales, while coal seams are developed in some cyclic sequences. Brackish-water or marine sediments sometimes interfinger with these rocks, indicating that they were deposited in shallow, marginal basins. In some zones, however, marine rocks dominate: thus the Middle Miocene rocks found along the northern edge of the Alpine–Carpathian belt typically are marine in the area east of Munich, and they form the greater part of the molasse sequences in the eastern Mediterranean and Adriatic. In complete contrast, elsewhere, for instance in the Balkans, Cyprus and what was Yugoslavia, evaporites accumulated in Miocene times.

Molasse, then, is typical of the later stages of the orogenic cycle. We can thus liken the Alpine molasse to the Old Red Sandstone, which was laid down in the post-orogenic phase of the Caledonian cycle. Similar deposits have been found to occur among the sequences of most orogenic cycles and, like flysch, are useful marker horizons that allow geologists to establish what stage in the orogeny has been reached at a particular point in the succession.

The European craton

During the late Palaeozoic and early Mesozoic times, the climate appears to have been hot and arid over much of the region. Strata of Permo-Triassic age are widespread and these include fossil sand dunes, as well as other typical continental deposits laid down in rivers and lakes that periodically dried up. The reddish colouration such rocks acquire has led to them being called New Red Sandstones. Not all of the region was desert, however, and there is clear evidence that shallow seas covered parts of the continent from time to time.

Marine incursions inundated parts of both central and northern Europe. The thick series of dark bituminous shales (called Kupferschiefer), evidently accumulated under a body of standing water during the late Permian; this body was called the Zechstein Sea. The Kupferschiefer or 'copper shales' are of particular importance since they became impregnated with copper and provided Germany with an important source of copper ore. Another important commodity won from these rocks is salt; important evaporites were laid down and the shallow sea repeatedly dried up under the scorching sun. Indeed, the saline concentration became so high that the teeming life which had characterized its waters was wiped out as thick beds of rock salt (anhydrite), potash and magnesium salts were laid down. The famous Stassfurt Mines have been carved into these rocks.

Much of the salt rose as diapirs (tear-shaped bodies) into the Mesozoic and Tertiary strata which overlie the craton and which lie beneath the Low Countries and southern North Sea also. Salt is less dense than surrounding rocks, and has the ability to force its way upwards, rising as diapirs that slowly

Liassic limestones and mudstones at Harbury, Warwickshire, England. Such rocks are typical of the shallow epicontinental sea that spread across the European craton. Rich in marine fossils, particularly ammonites, they attract fossil collectors in large numbers.

White chalk cliffs with flint layers, Denmark. Chalk, a pure organic limestone, was laid down over vast areas of America and Europe during the Cenomanian transgression. This inundated both cratons and was undoubtedly the most widespread event of its kind in the geological record.

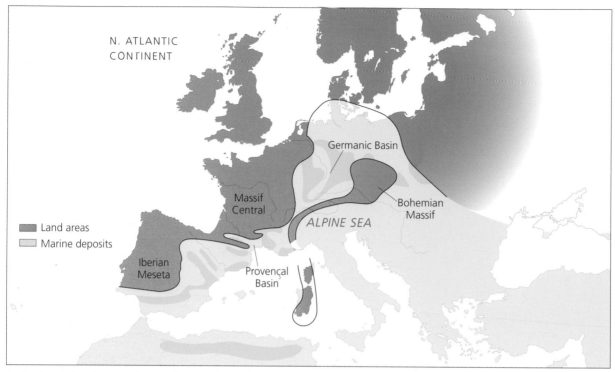

The Muschelkalk Sea which spread over the European deserts in Triassic times. Its animal population has a number of special characteristics – it was rich in individuals but poor in species – and seems to have derived from the deeper Alpine Sea to the south. Land areas are shown in brown, marine deposits in green.

shoulder the country rocks aside. The salt domes provide convenient traps for hydrocarbons. The Zechstein Sea evidently did not connect with the Tethys Ocean, but was actually an arm of the Arctic Ocean.

Somewhat later, during mid-Triassic times, the craton sat to the north of the relatively deep Tethys Ocean which lay between North Africa, southern Europe and southern Russia. The continent was the almost peneplaned remnant of the ancient Hercynian Mountains and the older cratonic core, which had been transformed into a great desert. During the Jurassic period, the sea invaded the southern border a number of times; because it was of such low relief there was little coarse detritus available, and the characteristic sediments were calcareous muds (called marls) and organic limestones. These collected in large basins, one of which is known as the Anglo-Parisian Basin – currently bisected by the English Channel and joined by Eurotunnel! Others developed in Aquitaine, in northeastern Spain, over southern France and also in the Black Forest and Bohemia.

During the Cretaceous period there was a more widespread invasion of the sea and many of these Jurassic basins were blanketed in marine Cretaceous deposits. Indeed, during the early part of Late Cretaceous time, the sea flooded almost all of Europe and Russia south of the Baltic Shield and much of northern Africa. It also invaded North America, lapping up against the Appalachians and spreading over the continental interior. It was in the Phanerozoic era that the Sahara was inundated by sea-water for the first and last time. This was one of the greatest marine transgressions of all time. Beneath its waters accumulated the familiar white chalk, a soft white limestone which built the White Cliffs of Dover. The chalk is composed of millions of tiny algal skeletons that flourished in the warm Cenomanian seas.

At the beginning of Tertiary time central Europe was low-lying. The Baltic Shield, to which the Caledonides of Britain and Scandinavia were by now strongly welded, behaved as a positive region, but the old Hercynides were much fractured, buried by sediments and only parts of them stood out above the general low level. About 60 million years ago, the sea

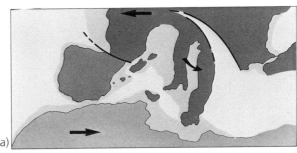
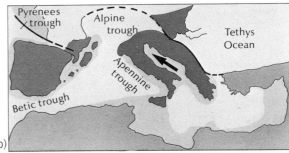
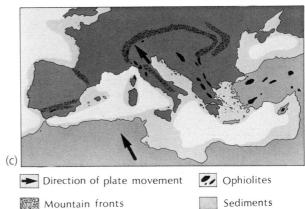

slowly expanded over a vast area, spreading westwards until by late Oligocene times (about 25 million years BP) it had reached southern Russia. Later, during the Miocene, it slowly withdrew again, and by the late Pliocene apparently had departed the continental interior altogether.

The crust in upheaval

The cause of the Alpine orogeny, as we have seen, was the head-on confrontation between Eurasia and Africa. Plate movements gradually eliminated the intervening Tethys Ocean and brought together thick sedimentary belts that had lain both north and south of the line of contact. The movements also involved the underlying crystalline basement.

Although this brief description explains the broad features of the Alpine chain, it does not really answer all of the geological questions posed by the rocks. It has been found necessary to consider the possibility that, as well as the two major lithospheric plates, there were also a number of smaller plates (microplates) that became mixed up in the movements, but behaved independently of the major plates.

The Mediterranean poses a particularly knotty problem for interpreters of palaeogeography; here the arrangement of mobile zones is very complex and geophysical data suggest an unusual story. Palaeomagnetic data from rocks ranging in age from Palaeozoic to Triassic show that Corsica, Sardinia, northern Italy and Iberia have undergone large rotational movements with respect to the main European craton. The Italian peninsula, for instance, seems to have been rotated about 43° clockwise during the late Palaeocene and 25° anticlockwise since the mid-Eocene!

High among the fold mountains of southern Spain there are detached folds – nappes – which must have come from a source region currently to the south and where now there is deep water. Then again, amidst

The development of the western Mediterranean during the Alpine cycle. (a) During Middle Jurassic time (165 million years BP) the African and American plates separated, pushing Africa east with respect to Europe. Small microplates set between the main plates were rotated anticlockwise as a result. (b) Subsequently, between 80 and 40 million years BP, the separation of plates bearing North America and Europe reversed the relative movements between the European plate and the smaller plates to the south. (c) More recently, the African plate drifted northward, completing the formation of the Alpine chain, and giving us the modern configuration of this region.

the Italian Apennines are clastic sedimentary rocks that derived from land somewhere southwest of the present Italian peninsula. Difficult though this is to believe, we have to accept that Sardinia and Corsica once occupied this position, and that the Tyrrhenian Sea – which currently separates Italy from both Corsica and Sardinia – opened due to the clockwise rotation of Italy.

Although of the same age, the Pyrenees do not form a part of the Alpine chain; in fact, they appear to have originated in a tremendous rucking up of

(a)

(b)

Aerial views of part of European Alps showing: (a) glaciated valleys, sharply-defined aretes and cirques partially filled with ice and snow; (b) folded metasedimentary rocks.

Recumbent folding in the rocks of the French Pyrenees.

strata which overlaid the ancient craton. This was followed by gravity sliding of the rocks, one over the other, as movements occurred in the underlying Hercynian basement. The more central regions of this range seem to have been elevated at least several kilometres, while the basins to the north and south were filled with both flysch and molasse of Tertiary age.

As we have seen, the Apennines are also an integral part of the great Alpine chain, having been rotated from a position formerly much further west. Recent work suggests that west of Rome, in Elba and in Corsica are ophiolites. Since these represent fragments of obducted oceanic crust, here is further evidence for elimination of an ocean as collision took place.

As the Alps finally rose due to the convergence of the two cratons, so the Mediterranean – once opened to both west and east – began to close. Closure first took place in the east during the Miocene, as Europe and Arabia collided. Then, a little later (perhaps in Late Miocene times) palaeontological evidence sug-

gests that the western end also became closed off. This resulted in sealing-off the ocean, increasing evaporation rates and consequent deposition of evaporites during the Messinian age (about 6 million years BP).

The Himalayas

The Alpine mobile belt passes through the Middle East and then swings northeastward and merges with Earth's greatest mountain range: the Himalayas. Tertiary sedimentary rocks from Turkey and Iran comprise a thick succession of clastics which were eroded from uplands on the northern border of the Tethys Ocean. To the south, these strata pass into carbonate rocks, deposited in shallow water. The main folding of these rocks was accomplished during Miocene times, when the Zagros Mountains of Iran were squeezed up as a result of collision between Eurasia and Arabia, then a part of the African plate.

The Himalayas themselves are part of a vast moun-

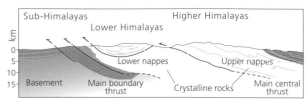

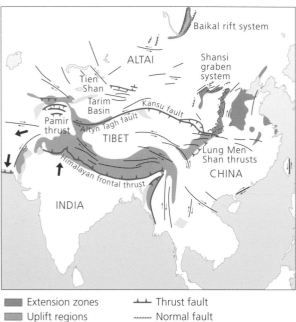

Extension zones — Thrust fault
Uplift regions .... Normal fault
Folds ═══ Strike-slip fault

Africa and Asia collide to produce the Himalayas and the Tibetan Plateau. Between 60 and 40 million years BP plates containing Eurasia and Peninsular India collided. India then slowed down but nevertheless continued to thrust northward for at least another 2000 km. The fact that buoyant continental crust does not subduct meant that some of the pressures were absorbed by the stacking up of vast overthrust slices of rock. More of the horizontal compression was taken up by the uprise of the Tibetan Plateau, while China was pushed forcibly eastward along enormously long strike-slip faults.

Landsat image of the collision zone between India and Asia. Mount Everest is located towards the top right-hand side. The Ganges River is towards the bottom.

tain and plateaux complex which occupies large tracts of central Asia, Mongolia and Tibet. There is at least 2 500 000 km² of terrain over 4 km above sea level in this huge massif. The scattered geophysical data that exist suggest that the continental crust is roughly twice its normal thickness here, i.e. 70 km instead of the usual 35 km.

Only the south and west parts of this belt are an integral part of Alpine movements. The ranges to both north and east are built mainly from faulted blocks of ancient Pre-Cambrian rocks and Palaeozoic orogenic belts which were affected by block movements in late Tertiary times. These motions specifically were linked with collisions between the Indian subcontinent and Eurasia, which began about 40 million years ago. After the initial contact India decelerated, but continued to drive northward into Eurasia for another 2000 km. It is not surprising the effects have been so marked!

The colliding plates were both continental in character and, since this kind of material does not readily

subduct, the pressures of collision had to be absorbed in some other way. Part of the stress was relieved by a stacking up of huge slices of rock which were scraped from the forward-driving front of the Indian craton. These slices – which moved roughly west to east – were translated along low-angled thrust planes, involving both marine rocks from the mobile zone and also basement materials. The sideways motion of these rocks was due at least in part to the fact that the converging plate margins were non-parallel, thus driving the first regions of contact in this direction. It is from these rocks that the Himalayas were formed.

A further part of the compression appears to have

been accommodated by the upfaulting of large tracts of crust, giving rise to major topographic highs such as the Tibetan Plateau. Even allowing for this, together these movements could account for only one half of the required 'give'. It has been suggested that much of the convergence was absorbed by China which was shoved eastwards, out of the way of India, along enormously long strike-slip faults.

With a little bit of ingenuity it is possible to discern the line of demarcation between the northern and southern plates. This is known by geologists as the Indus Suture since it follows the line of the River Indus in northeast Pakistan. It then crosses towards the Brahmaputra River in Bangladesh. The folded rocks produced during the final cataclysmic stages of the orogeny now lie to both the north and west of this line, forming the great ranges of the Pamirs, Karakoram amd Hindu Kush. To the south lie huge stacked slices of crust derived from the edge of the Indian craton. Beneath this emerges the undeformed continent of India, spreading southwards over the peninsular plains.

The geological cycle whose record we can recover in the northern ranges of the Himalayas took considerably longer to run its course than did the Alpine movements. There are no major tectonic breaks between the Proterozoic and late Mesozoic strata in this region. The Himalayan flysch deposits – characteristic of orogenic activity – span the interval between the mid-Cretaceous and Eocene, indicating that orogeny at the southern margin of the Eurasian continent began in late Mesozoic times, about 95 million years ago. The non-marine molasse was laid down significantly later than in the Alps, in late Miocene times, and during the Pliocene and Pleistocene. As in the Alpine region, these deposits partly post-date the major orogenic episode, yet even these youthful strata have been deformed and thrust beneath the southern Himalayas.

Events in Siberia

The Ural Mountains were periodically rejuvenated during both Mesozoic and Tertiary times and, as a consequence, expose ancient Pre-Cambrian and Palaeozoic rocks. The Urals were marginal to the craton which had become stabilized in Pre-Cambrian times and has, since that time, received a variety of shallow-water and continental sediments which, in the Verkhoyansk and Khakingat Basins, attain a thickness of over 6 km, spanning a time interval of 1500 million years.

During late Permian and early Triassic times there were extensive outpourings of basaltic lavas over the northwestern part of Siberia. This major igneous event had no parallel on the European craton, and rivalled in extent the Keweenawan lavas of North America. The lavas and their associated feeder dykes cover an area of around 1 500 000 km^2 and in places are 1 km thick. Basalt is not, of course, a characteristic rock of continental cratons and was extruded as a result of crustal stretching which allowed magma to rise upwards from the underlying mantle. Crustal fracturing is also suggested by the occurrence of diamond-bearing volcanic pipes further east, also of Triassic age.

During Jurassic times the sea invaded much of southern and eastern Siberia but the great Cenomanian Transgression, which deposited the Chalk in Europe, did not extend this far. Instead there was the deposition of immense coal deposits, presumably under conditions similar to those found in both Europe and North America in Carboniferous times.

Part Four

Gondwanaland and more recent events

(a) Close-up of ancient ripple marks in sediment formed on Gondwanaland in late Proterozoic times. (b) The same ancient ripples, seen in outcrop, Arkaroola, South Australia.

19

Gondwanaland before the break-up

The unbroken supercontinent

It is the end of the Palaeozoic era. The interior of Gondwanaland has become stabilized, experiencing little tectonic disturbance. The Equator crosses through the northern end of the Red Sea (which does not yet exist), passes through the Sahara and thence through southern Florida. On the west and south a series of marginal basins runs along the western side of what now is South America, through western Antarctica and into eastern Australia. Along the perimeter, the southern margin of the Tethys Ocean runs through what is now North America, Arabia and northern India.

I have already outlined the Late Palaeozoic history of this great supercontinent, describing how during Pennsylvanian and Permian times there was widespread glaciation of the interior. Many of the glacial beds are overlain by Permian continental deposits, and incursions of the sea are indicated by occasional marine horizons. The accumulation of continental deposits continued in some parts of Gondwanaland right through into early Cretaceous times, a period of around 150 million years. This shows just how stable the continent was. Laurasia and Gondwanaland were joined during the Triassic, the northern coast of South America and eastern Central America being in contact, as were northwest Africa and the eastern margin of North America, Spain and North Africa.

The rocks of Gondwana

Predominantly non-marine sediments accumulated in several of Gondwana's basins during late Palaeozoic – Mesozoic time. Occasional marine incursions are indicated by the interbedding of marine bands which have proved extremely useful in correlating strata from one region to another. The sequences found on the various cratons that were then an integral part of the supercontinent show remarkable similarities, as we might expect if they

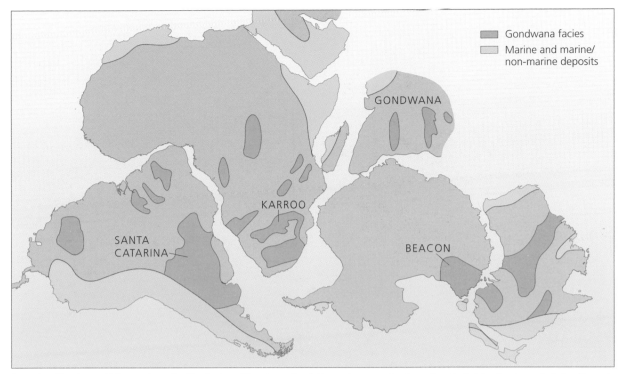

The rocks of Gondwana. This map shows the distribution of Permian and early Mesozoic non-marine formations (called the Gondwana facies) on the various components of Gondwanaland. The accumulation of continental sediments started during the late palaeozoic and continued in many places until the Jurassic or even Cretaceous periods. Thus, in some basins there is a record of continuous non-marine sedimentation spanning 150 million years.

were all close to one another.

The cratonic sequences span the period from the Late Carboniferous (Pennsylvanian) to the Cretaceous. The lowest strata of the Gondwanaland succession are the beds I have already described which are glacial deposits. These are the ones which suggest that much of the supercontinent was at high latitudes at this time. The overlying Permian sequence includes coal horizons that contain the Glossopteris flora. Higher in the sequence these plant-bearing strata give way to yellow, green and brown sandstones and marls which contain abundant reptilian remains. Study of these rocks and their fossil content indicates that Gondwanaland experienced a gradual amelioration in climate, reaching a peak in Late Triassic times, about 215 million years ago.

Those parts of the supercontinent which now comprise Africa and South America appear to have slowly drifted away from the polar regions towards the Tropics. The Triassic sandstones, so typical of all the basins, were parallelled by similar aeolian rocks – the New Red Sandstone – in Laurasia. This was then located within northern tropical latitudes.

Shales, fine-grained sandstones, aluvial deposits and some limestones accumulated in the cratonic basins, in shallow water. The rate of sedimentation would have been very finely balanced against the rate at which the basins subsided, ensuring that the depth of water did not change much over long periods. Of the one or two marine incursions that did occur, one particular band, of Early Permian age, was rich in the remains of the marine shell *Eurydesma*, an indicator of cool marine conditions. This horizon, known as the Eurydesma Band, is found in both southwest Africa and Brazil, and has proved useful in correlation between the two continents.

Mesozoic life

The later part of the Mesozoic era was a time of considerable biological change. Many of the invertebrate animal groups that had flourished during the earlier Mesozoic had disappeared altogether by the close. The nearest example of this is perhaps that of the ammonites: they became extremely prolific during

A late Jurassic landscape, showing the gigantic sauropod dinosaur, *Brontosaurus*, which must have weighed over 30 tons when alive.

the Jurassic and Cretaceous, then slowly declined. Plants, too, underwent significant changes, the greatest being the evolution of flowering plants (angiosperms). These first made their appearance in the Early Cretaceous and, by the Late Cretaceous, had taken over from their earlier cousins, the gymnosperms – of which the modern conifers are the descendants. Fish also underwent changes: the earlier Mesozoic forms still had heavy scales and skeletons built mainly from cartilage. By the close of the era, small scales and bony skeletons were the rule, as they are in modern fishes. The amphibians continued to decline, and the reptiles saw dramatic changes, best exemplified by the dinosaurs.

During the first part of the Triassic period there was a wealth of reptilian life on the continent, from one group of which – the thecodonts – were descended the dinosaurs and, later, the birds and crocodiles. These thecodonts and their relatives ruled the land for over 150 million years, starting out as small lizard-like animals, then developing into some of the most massive heavy-tailed animals that could walk on two legs. The more primitive forms seem to have been cold-blooded, like living reptiles; however, detailed physiological studies of their remains have suggested that some of their later descendants were warm blooded, and presumably were more active

than once thought.

The more advanced dinosaurs can be divided into two groups: those whose bone structure points to a reptilian mode of travel (saurischian or 'lizard-hipped'), and those that were 'bird-hipped' (ornithischians). Both appear to have been warm-blooded and there is some evidence suggesting that many were capable of quite rapid movement. One of their more pronounced evolutionary trends was toward gigantism: some were real monsters, standing 15 m high and long. Not all, of course, were this large; some were more dog-like in proportions. All had extremely small brains for their size.

The immense forms we have all seen depicted in popular books and films were descended from the saurischian branch. *Brontosaurus* and *Diplodocus* are but two of these monster forms; some must have been amongst the fiercest carnivores on Earth. Ornithischians, on the other hand, were mainly herbivores, but also grew to large size. The armoured *Stegosaurus* is probably the best known of this group; it flourished during Jurassic times. Another member, *Triceratops*, had strong horns with which presumably it could attack its enemies with some vigour. This form and its close relatives roamed about in large numbers during the Late Cretaceous but by the close of that period were extinct.

Some dinosaurs also lived in marine environments, while others learned to fly. The latter were, of course, the precursors of modern birds, although the exact evolutionary link is rather obscure. Mammals also evolved during the Mesozoic, probably developing from reptiles during the Triassic. They may well have had a diet of insects and small reptiles.

As far as plants are concerned – palm-like cycads achieved prominence during the Jurassic and early Cretaceous, while the seed-ferns declined. Conifers and gingkos proliferated, and the flowering plants made their first appearance. The earliest of these seems to have appeared during early Cretaceous time.

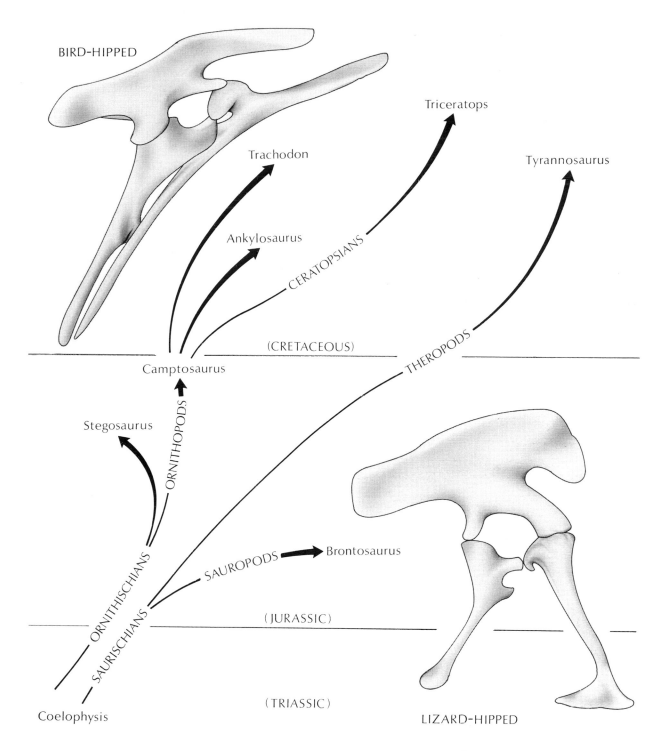

BIRD-HIPPED

Triceratops

Trachodon

Ankylosaurus

CERATOPSIANS

Tyrannosaurus

(CRETACEOUS)

Camptosaurus

THEROPODS

Stegosaurus

ORNITHOPODS

ORNITHISCHIANS

SAUROPODS → Brontosaurus

(JURASSIC)

SAURISCHIANS

(TRIASSIC)

Coelophysis

LIZARD-HIPPED

Diagram showing the two orders of dinosaurs. These developed from a common ancestor but had distinctly different pelvic structures. The ornithischian branch ('bird-hipped') differed from the saurischian ('lizard-hipped') in the arrangement of the hip structure.

These flowering plants underwent their greatest proliferation during the Jurassic and Early Cretaceous, just as the ruling reptiles were giving way to the mammals. The causes of these dramatic changes are still being debated. I shall return to this topic later.

The dismemberment of Gondwanaland may have begun with the generation of domes above lithospheric heat sources (hot spots). Such domes often cause the crust to open along a pattern of three rifts in the pattern of a three-pointed star. Two rifts of each group ultimately connect with their neighbours to form the suture along which the new ocean grows.

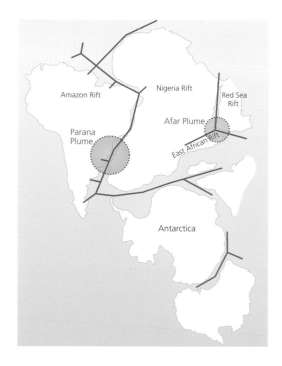

The beginning of the end

Before the close of the Eocene epoch, not only Pangaea but Gondwanaland were to fragment entirely. The precursor to these end events was the outpouring of vast amounts of basalt lava – now existing as what is termed 'plateau basalt' – largely during the Jurassic period.

There is ample evidence to support the idea that, during the Triassic (around 200 million years BP), a plate carrying Gondwanaland began to separate from one bearing Laurasia. At the same time as sediments of Late-Triassic age were being laid down in fault-bounded basins that had developed along the east coast of North America and northwest Africa, basalt lavas were rising towards the surface in both regions. These lavas are radiometrically dated to 192–202 million years BP, which corresponds to the period spanning the Late Triassic to Early Jurassic. At roughly the same time, rather more alkaline lavas were being emplaced in North America. The Late Mesozoic was altogether a time of considerable igneous activity: such would be expected to accompany vigorous mantle and plate activity.

If we take a look at the rocks of the ocean floors, we find that the oldest known sediments to be deposited on the oceanic crust that eventually developed to separate North America from Africa were recovered from near the Bahamas Bank by the oceanographic research vessel *Glomar Challenger*. These proved to range in age between 162 and 151 million years BP (Late Jurassic); similar rocks were found some 530 km from the margin of the North American continent, east of North Carolina; these yielded an age of 155 milion years. Knowing the distance between two magnetic anomalies recorded in the sea-floor basalts (one dated at 153 million years and the other at 107 million years), geophysicists estimated that between 153 and 107 million years ago the sea-floor spread at an average rate of 1.1 cm per year. Using the simple formula: *distance = spreading rate × time*, it can be seen

that the time that must have elapsed since the spreading began equals 530 divided by 11 (11 being the distance in kilometres travelled by the sea-floor in one million years), giving an answer of 48 million years. This puts the time of initial separation at about 202 million years BP.

Palaeomagnetic data show that the oldest magnetic anomaly on the floor of the central Atlantic – then separating North America from Africa – has an age of 153 million years. This anomaly is located 535 km from the eastern margin of North America, so, again assuming a 1.1-cm-per-year spreading rate, the two continents would have commenced separation about 202 million years ago.

The *Glomar Challenger* also collected deep-sea cores of ocean sediments from near the centre of the Gulf of Mexico which, at the time of separation, would have been on the line along which initial spreading would have commenced; these proved to be of Late Jurassic age, proving that separation must have begun well before this time.

Studies made of the way in which continents have moved with respect to the magnetic poles also add weight to the view that break-up started in Early Jurassic times. 'Polar wandering' curves plotted for Africa and North America agree well prior to Carboniferous–Triassic time, but for the Jurassic and thereafter, they begin to diverge more and more, showing that the continents went their separate ways from this time onwards. The situation between North Africa and Europe is less clear. They must have been separating by at least 180 million years BP, some geologists preferring to put the time of initial separation even earlier. My own preference is for the Late Triassic. Either way, this began the final slow fragmentation of Pangaea, supercontinent extraordinary.

The plateau lavas

Just before the break-up occurred, there was a significant episode during which enormous volumes of fluid, basaltic magma was poured out. These lavas emerged largely from fissures in the cratonic crust which opened in response to mantle-driven lithospheric stretching. Sills and dykes of dolerite – the coarser-grained equivalent of basalt – filled countless

fissures, many of which extended well beyond the limits of the flows. Because they spread out over extensive regions, forming flat-lying topography which has now been etched by erosion, these are known as *plateau lavas*.

In northern India such lavas formed the extensive Deccan Traps, a sequence that reaches 3 km in thickness and extends over 550 000 km². These were extruded during late Cretaceous and early Cenozoic times. Similar rocks are found in Brazil, where an at least 6 km thickness of volcanic and non-marine sedimentary rocks of late Cretaceous age accumulated. Plateau basalts occur also in southern Africa, where the Upper Gondwana Basalts cover large areas of the eastern region; there are also countless related sills and dykes over a much wider area of the old craton. Interbedded with the flows are sediments containing Jurassic plant remains; radiometric dating confirms this age.

Stepped topography in horizontal sequence of plateau basalts, Gib-Bheann, Mull, Scotland.

Also featuring in the Africa story are *kimberlites* – rocks that are found in narrow volcanic pipes and which appear to have formed in response to explosive volcanism from very deep levels. Kimberlites contain olivine, pyroxene, mica and garnet and are rich in the element magnesium. They include large numbers of fragments of other rock types, most of which appear to have been formed at very high pressure, presumably deep within the lower crust or upper mantle. It is within these pipes that many diamonds are found.

The principal basaltic activity seems to have been concentrated in Jurassic–Cretaceous times, but the basalts are not the only products of this active period of Earth history. Other igneous rocks which are particularly rich in the alkali elements potassium and sodium, have been found on all of the southern continents, save for Antarctica. These are usually coarsely crystallized and occur in ring-shaped masses, many of which appear to be part of larger volcanic centres. Such rocks are typical of regions of continental crust which are thinner than normal, generally due to rifting. The rocks are of comparable age to the plateau basalts.

The igneous activity which was a feature of all of the Gondwananan cratons during Jurassic–Cretaceous times marked a period of crustal stretching and eventual rupture of the old supercontinent along a number of rift zones. These eventually became active spreading centres from which the new oceans that were to separate the modern continents developed. They are still active today; the continents continue to drift and the pattern of land and sea is slowly, but constantly, changing.

20

Gondwanaland disrupted

The quiet period ends

The intense magmatic activity which marked the outpouring of the Plateau Basalts signalled the end of Gondwanaland as a supercontinent. So much basalt was erupted that those parts of the crust that experienced stretching were also depressed by the lava overburden. After continental drift had begun, naturally, these regions became the peripheral areas of the new continents; it was here that a new generation of sedimentary basins developed. Initially these were shallow and, having been invaded by shallow water, the marginal seas often underwent evaporation in the lower latitudes. In consequence some of the earliest deposits of these new basins were evaporites; indeed, much of the salt was initially deposited in down-faulted graben. Gradually the marginal basins subsided more and more and the salt water of the Tethys Ocean invaded them. The situation would at first have been rather like the present day Red Sea. In time, however, as drifting continued and the basins developed, the modern Indian and Atlantic Oceans grew.

The southward expansion of Tethys can first be traced as it passed between eastern Africa and India, about 160 million years BP. Gradually the waters linked with those already in existence in the rifts around Madagascar and South Africa. Ten million years later, the ocean also spread up the eastern coast of Peninsular India. The ocean began to separate South America and Africa at some time between the Late Jurassic and Early Cretaceous, i.e. 150–130 million years ago. More recently – about 80 million years BP – the ocean came between New Zealand and Australia, which was previously still attached to the western edge of Antarctica. The rifting apart of this craton took place 40 million years later. The expansion of the new oceans and oceanic crust, and the movement of the continents, have continued until the present day.

One might reasonable enquire why the break-up

Horizontal basalt flows rifted apart during the break-up of Gondwanaland. Near Manakha, Yemen.

occurred. Many geologists would point to renewed activity in the mantle. However, I must take time to mention briefly a very 'way-out' idea which was put forward during 1996. Michael Rampino – professor of Earth Science at New York University – and Verne Overbeck – a scientist at NASA's Ames Research centre in California – suggested that a giant 15-km-diameter asteroid crashed into the earth, creating fractures in the planet's crust which led to the formation of today's continents. The shock waves sent out by the event were supposed to have created the triangular shapes of present-day South America, southern Africa, India and western Antarctica. They claimed to have found a 300-km-wide undersea crater off the coast of the Falkland Islands which was the result of the collision.

When the idea was first proposed, not surprisingly, it was greeted with incredulity by the scientific community. It has to be said that there is much scepticism still, and most scientists would prefer to cite some internal process. Nevertheless, one should keep an open mind.

The East African Rift Valley

One of the main physiographic features of Africa is the Great Rift Valley. Lying well to the east of the axis of the ancient craton, this feature is the major part of a 5000-km-long tract of fracturing that extends from the Limpopo River in the south, to the Red Sea and the Jordan Valley in the north. Along its path it crosses the great crustal dome which covers much of Kenya. The southern end of the rift appears to have opened during the Mid-Tertiary, but the main phase of faulting associated with the doming of the craton by a rising mantle plume spanned the Miocene–Pliocene and was accompanied by volcanism and intrusion. The rift is a region of present-day high heat flow, and subject both to seismic and volcanic acitivity. Vertical displacements along the main fault – which is a graben – are considerable: in the regions of Lake Nyasa and Tanzania the floor lies between 2 km and 3 km below the rim; spectacular fault scarps are to be seen.

While there are modest accumulations of Mesozoic and younger sedimentary rocks on the rift

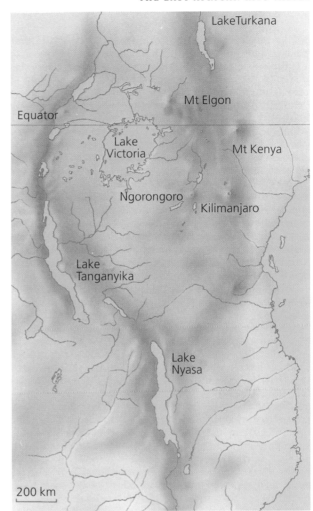

floor, it is the volcanic rocks which dominate the stratigraphy. Igneous activity commenced in the south during the Mesozoic and then spread northwards; it continues until the present day. The main rift splits into western and eastern branches, the former running through Uganda, the latter through Tanzania and Kenya. A host of explosive volcanic centres pierce the floor of the western branch, the magma being enriched in alkali elements, particularly potash. Peculiar lavas called carbonatites are typical in this region. These are very unusual in being composed almost entirely of sodium carbonate. The very explosive character of some volcanoes indicates that volcanic gases play a vital part in rift magmatism, coupled with relatively high viscosity for some magmas, possibly due to mixing of mantle and continental materials.

The floor of the eastern branch, together with the rift borders, is flooded by plateau lavas. These are particularly evident in Ethiopia and northern Kenya, and include basalts, phonolites and trachytes. The huge central volcanoes of Mounts Kenya, Kilimanjaro and Meru lie on the rift shoulder, further south. Their lava chemistries are similar, alkali elements being characteristically abundant. Some of the world's most stunning scenery is to be seen in this fascinating region.

The East African Rift Valley. This huge downfaulted region extends through the eastern side of Africa from the valley of the Limpopo River, through Ethiopia, to the Red Sea–Jordan Valley fracture belt. It has a total length of about 5000 km.

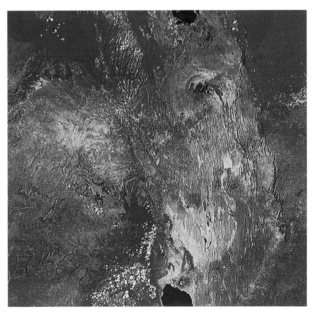

Landsat image of a part of the East African Rift in southern Kenya. Lake Natron, situated on the border between Kenya and Tanzania, can be seen at the bottom of the picture. In this false-colour image, vegetation is picked out in red hues, and is concentrated on the higher ground. Nairobi is located near the eastern edge of the frame, just above centre. Dominating the picture is the rift faulting which here runs north to south. A number of volcanoes can be seen on the rift's floor, particularly obvious being Suswa, situated west of Nairobi and south of Lake Naivasha (top).

View of the rift floor with volcanic cones. In the distance is the opposite wall of the rift, composed of plateau basalt lavas. View from Mount Meru, Northern Tanzania.

Sequence of phonolite and nephelinite lavas in the caldera wall of Mount Meru volcano, northern Tanzania. A number of very large, recently-active, strato-volcanoes – including Kilimanjaro and Meru – sit near the shoulder of the main rift fault.

Marginal basins

During Jurassic, Cretaceous and Tertiary times, marine sediments collected in large basins along the perimeters of each of the ancient cratons that collectively had formed Gondwanaland. These sediment-filled basins are what geologists technically term *aulacogens*. In the sequences now built up on the cratons, towards the interiors the basin sediments interfinger with shelf or continental rocks. The rocks which actually collected in the basins have proved immensely important since they are often invaluable hydrocarbon reservoirs.

In many cases the new basins ran along the sites of more ancient mobile zones, some of which formed as long ago as the Late Pre-Cambrian. This relationship is very clearly seen along the Atlantic coast of Africa, where a number of large basin sequences are to be found. One of these – the Senegal Basin – overlies the ancient Mauritanide mobile belt which had been active until the Late Palaeozoic. During Cretaceous and Tertiary times, marine sediments again collected here, the fill becoming 5 km thick! Many of the rocks are limestones and the Cretaceous strata are pierced by large salt domes. This reveals that evaporites were among the earliest basin deposits, a feature common to most new basins which formed at the appropriate latitudes.

Another sediment trap, the Benue Basin, extends inland from the Niger Delta. This has a width of at least 200 km and is filled with a succession of Cretaceous and Tertiary shallow-water sediments 6 km thick. Amongst the basement rocks to the northwest of the basin is a series of alkaline granites and volcanics highly enriched in tin. Similar enrichment occurs in a line of northeast–southwest-trending intrusions that pierce the crust on the opposite side of the basin. Significantly, these rocks extend seawards into the chain of volcanoes that ends in the still-active São Tomé centre. This is just one illustration of the close connection which exists between vol-

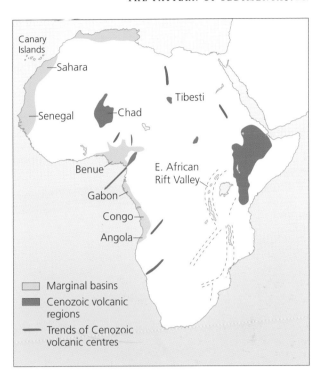

The marginal basins of the African continent. Many of these formed in response to subsidence of the edges of the craton along ancient fractures. During Cretaceous and Tertiary times substantial accumulations of sediments collected in them. Similar basins developed at the edges of the other continents.

canism and these sites of early continental rifting.

Large basins also occur in Zaire and Angola, along the east coast and within Arabia, and along the north coast of Africa, where Jurassic strata – which include many limestones – were laid down in a shallow sea that must have connected with the Tethys Ocean to the north. Peripheral basins of this kind are found also on the other continents and are of comparable age and size. Africa simply provides us with a particularly good example of Late Mesozoic and Tertiary developments.

The pattern of sedimentation
Africa

Marginal basins were a characteristic of the continental perimeters, but geological processes also affected the continental interiors. Again, I will use Africa as an example, for its Tertiary history is in many ways paralleled on other Gondwananan fragments. The later history of this continent is dominated by evolu-

Desert weathering of Cretaceous Khurma Sandstone, Petra, Jordan. These were sediments deposited in Gondwanan basins.

tion of the Earth's largest cratonic landmass, as well as the growth of the Rift System.

As I have already described, the north–south dome over which the rifting developed lies well to the east of the African axis. This arched region of ancient crust forms a major watershed, from which, to the east, the land falls simply away towards the coast. To the west, however, things are rather different: several major sags in the craton received non-marine sediments throughout the Palaeozoic. These interior basins, of which the Kalahari and Congo are the best known, were the collecting ground for a variety of lake sediments, wind-derived sandstones, and a kind of deeply-eroded calcareous dust called loess.

At the continental margin, terrestrial deposits interfinger with shallow-water marine rocks; these late Cretaceous–Early Eocene strata are found along the east coast of South Africa and also in Libya, Egypt and Saudi Arabia. To the northwest – in the Atlas Mountains and also in the Elburz and Zagros massifs – calcareous sediments are often accompanied by evaporites which pass upward into wedges of clastic rock worn from the newly-elevated lands to the south. The

huge oil reserves of the Middle East are found mainly within the Tertiary limestones on the southern flanks of the Zagros Mountains of Iran.

More recently, the geological evolution of Africa has been dominated by epeirogenic (up-and-down) movements and associated periods of erosion, whereupon huge tracts of horizontal or subhorizontal strata have been hewn into major erosion surfaces that have been deeply dissected by rivers.

South America

A similar story can be written here. The majestic Andes eventually were to rise along the sites of marginal basins that continued to be uplifted during the Tertiary. The most ancient rocks – those of the Guyanan and Brazilian Shields – are situated in the broadest part of the craton, and are separated from one another by the huge Amazon Basin which has collected not only recent alluvium but also sediments dating back to Palaeozoic times.

During the Cretaceous period the sea along the western margin of South America was quite restricted but, because of the continued rise of the

Volcanic landscape of the Warrumbungles National Park, Coonabarabran, New South Wales, Australia. Located on the eastern side of the craton, this was within a zone of intense geological activity at this time.

Andean chain, this was even more marked during the Tertiary period. In consequence by Palaeocene times, marine deposits became restricted to a narrow coastal belt running along the western perimeter. At this time, remember, South America was in the process of breaking away from North America; in Eocene times the ocean breached the Andean barrier and flooded an area east of the Andean highlands, forming a shallow epicontinental sea. The main Andean orogeny occurred during the Mid-Miocene (between 17 and 15 million years BP), since which time marine sedimentation has been confined to the extreme western margin of the continent. The continental plains which now separate the ancient shields from the Andes are largely covered by clastic material eroded from the great Andean mountains.

In the far north a thick wedge of largely marine Cretaceous and younger sedimentary rocks lie adjacent to the Caribbean orogenic belt. These have acted as reservoirs for hydrocarbons, hence the very important oilfields that extend from Venezuela to Trinidad.

Australia

By late Mesozoic times, the upheavals associated with the Tasman orogenic belt were all but over, and the craton subsequently was warped and distorted but never folded. Very vigorous volcanism occurred in several places. The Murray Basin – which is bounded by the Flinders and Mount Lofty Ranges – together with the Carpenteria, Lake Eyre and Darling-Warrego Basins – collected continental sediments that range in age from Late Cretaceous to the present day. These include coal deposits. Marine rocks are restricted to basins that developed along the continent's perimeter during Tertiary times; many of these are carbonates. During Palaeocene times, however, the sea inundated the interior and reached a maximum extent during the Late Eocene. Subsequently it retreated and continued to do so until the end of the Tertiary period.

In contrast to the relatively stable conditions on the craton, there was considerable activity in the east. Indeed, the Eastern Highlands have been very recently active. Uplift of this tract began in Early

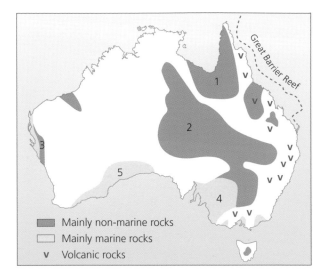

Map showing the Tertiary basins of Australia: (1) Carpentaria; (2) Eyre; (3) Carnarvon; (4) Murray; (5) Eucla.

Mainly non-marine rocks
Mainly marine rocks
v Volcanic rocks

Tertiary times and peaked in the Miocene epoch. Minor pulses of activity have continued into the Pleistocene. The Miocene disturbances were enough to drive out the sea from the marginal basins. Volcanism occurred during Palaeocene and Eocene times, along a broad belt extending from Victoria, through Queensland to Tasmania. The rocks, which include alkali-basalts, nepheline-bearing lavas and trachytes, were almost certainly erupted when Australia and Antarctica began to drift apart, about 53 million years ago.

New Zealand
This was a region of low relief at the beginning of the Tertiary and was inundated by the ocean during Eocene and Oligocene times. In later Oligocene times the sea retreated once more for which reason Tertiary sedimentation was dominated by marine clastic rocks, interbedded with limestones and coals. Andesitic lavas are abundant in the rock succession and the most acceptable explanation is that New Zealand was an island arc throughout most of Tertiary time.

India
While the recent geological history of India is dominated by the rise of the Himalayas, there was another Tertiary event of great significance. During the earlier Tertiary period, an important phase saw the voluminous outpouring of basaltic lavas, centred on the

Deccan region north of Bombay, giving rise to the Deccan Traps, a vast region of plateau flows which eventually accumulated to form a pile 3 km thick. These fluid flows spread out over extensive areas and were almost certainly associated with the separation of the Seychelles Islands from Peninsular India, about 70 million years ago.

Substantial continental deposits worn from the Himalayas have been dumped onto the craton and have continued to accumulate just south of the mountain range since Mid-Eocene times. This has contributed to the infilling of the alluvial basins of the North Indian Plains and the building of the huge deltas of the Indus, Ganges and Brahmputra rivers. Elsewhere on the craton, continental deposits are thin. Marine deposition did occur, but was confined to basins located on the margins which collected Tertiary sediments which, in the Cambay Basin, are between 2000 and 3000 m in thickness and oil bearing.

Antarctica
Antarctica's recent geological history is, to say the least, rather sketchy. However, in recent years geologists have been probing this ice-covered continent with geophysical instruments and now have a far better idea of what lies beneath the ice than, say, fifteen years ago. In general terms the story is not dissimilar to that of South America. Tertiary sedimentary and volcanic rocks accumulated in basins confined to the Pacific margin, largely between South Victoria Land and the Palmer Peninsula. Volcanism and the orogenesis associated with it were probably related to the eastward drift of oceanic lithosphere beneath the craton.

Life and death
During Jurassic and Early Cretaceous times the terrestrial landscape would have been dominated by seed-bearing plants (gymnosperms) and reptiles. Marine

Map of Yucatan, showing the site of the Chicxulub impact crater. 1–4 indicate successive rings (from gravity data).

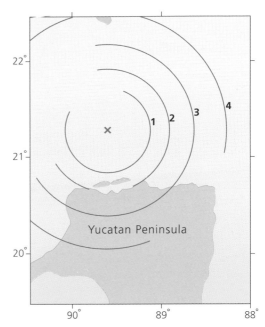

environments were home to plankton, molluscs, ammonites, corals, fishes, echinoids, crinoids and a variety of other shell creatures. Crustaceans lived in nearshore waters while, later in the Cretaceous, the more advanced angiosperms joined the plant kingdom and have persisted until the present day. Among their flowers and branches fluttered the first butterflies.

There is no doubt that life was very diverse at the end of the Mesozoic; however, not all organisms enjoyed the conditions in which they were living at this time, and some had been showing signs of slow decline. Indeed, at the close of the Mesozoic a large number of hitherto successful animal groups died out, including the dinosaurs. Many marine invertebrates also disappeared prior to the end of the Mesozoic, the ammonites being a case in point, while reef-building corals also perished. Others, however, survived, apparently unaffected by the changes which occurred, for example the crocodiles, lizards, turtles and snakes survived. Many theories have been forwarded in an effort to explain this period of *mass extinction*, but as yet no single explanation has struck an accord amongst scientists.

The death of the dinosaurs and all that

There is little doubt that an event or series of events occurred towards the close of the Mesozoic to bring about a marked change in the balance of species and genera which existed on the Earth. Apparently, about 65 million years ago, something affected the Earth in such a way as to cause the demise of a large number of hitherto highly successful animal groups. Indeed, it is not going too far to say that had not one of these – the dinosaurs – been rendered extinct at that time, the human species may not have become the 'supreme' product of evolution. Somehow the dinosaurs – which had been unchallenged for a very long time – died out, leaving the way clear for evolution of

more agile and intelligent mammals. What happened?

During the latter part of Mesozoic time the balance of evidence from stratigraphic and palaeontological sources is suggestive of a slow cooling of the world's climate. This in itself might have changed the balance of species, modified the food chain or driven some species along the path towards eventual extinction, but would not of itself have been expected to bring about the devastation which appears to have been wrought on so many groups and families.

The extinction of the dinosaurs has attracted the most press, not all of it very well reasoned, it has to be said. Certainly their demise occccurred pretty suddenly on the geological time scale; however, it certainly did not happen overnight; other families went far quicker, as we shall see. In fact – and this is not always mentioned in the more spectacular pieces of writing about them – they had been showing signs of slow decline prior for quite some time, at least 5–10 million years. Even so their abrupt departure at the K/T boundary was somewhat surprising, considering that the dinosaurs had been anything but an unsuccessful biological experiment. In a sense they had ruled the continental regions of the Earth for a great deal longer than mankind has done (and the way he is going, their record may yet stand!). But what about other groups which make up an equally vital part of the whole story?

The K/T boundary, Namibia, South Africa. [Photo: Patrick Moore]

The events – whatever they were – which led to the end-Cretaceous *mass extinction* affected land animals most savagely, but they also did much damage to the marine faunas of the Tethys Ocean – though it has to be said that families of most shallow-water marine invertebrates persisted into the Cenozoic. Heavy tolls were taken of some heavy-shelled bivalves, reef-building corals, many molluscs and, most of all, pelagic plankton; all disappeared simultaneously at what is called the K/T boundary (Cretaceous/Tertiary boundary). This 'moment of change' is represented by a distinctive clay horizon in many places; radiometric dates yield ages of 65.5 and 64.8 million years BP for the beginning and end of the period represented, i.e. a period of 700 000 years.

Such devastation could be caused by all sorts of things; for instance by a change in global temperature. This might have repercussions on the availability of vital food supplies, generate increased competition between certain species, and so on. However, these would be gradual rather than instantaneous in their effects. More rapid in their effects would be certain microbiological phenomena: it has been suggested that there may have been an upsurge in, or introduction of damaging viruses which would have had the most impact on species with the greatest number of individuals (e.g. the dinosaurs). These

are only some of the possibilities.

What evidence can we glean from the geological horizons which mark what is called the K/T boundary? Well, it is marked in Italy, Mexico and several other parts of the world by a rather distinctive layer of clay; this overlies freshwater deposits, and underlies a coal horizon. At the boundary itself, several tree species present when the freshwater beds were laid down were temporarily replaced by ferns until the former flora reappeared at the time the coal was laid down. Evidently something dramatic must have occurred to instigate this sharp change.

Significantly, within the clay horizon there is an unusually high content of the rare element iridium (150 times higher concentration than strata above and below) and also a high osmium isotopic ratio. While the former can be generated during major volcanic episodes on the Earth, this is not a common occurrence, and it has been suggested that the geochemical data can best be explained by a major asteroidal impact. Extraterrestrial bodies are known to carry elevated amounts of such elements and would, given the speed at which the final extinctions took place, provide a convenient explanation for the terminal events.

Since the asteroid catastrophe idea was first mooted there has been an avid search for the possible crater(s) formed during such an event. To date the strongest candidates are the 180-km-diameter Chicxulub Crater, located in the Gulf of Mexico, and the 32-km-diameter Manson Crater, in Iowa. These are of the same age and could be formed by, say, a 10-km-diameter asteroid which split into two as it tore through the Earth's atmosphere. On impact each bolide would have instantly vaporized, sending huge clouds of debris into the upper atmosphere. Such an event would not only cause almost total darkness for several weeks but also would have brought about a temporary change in climatic conditions and possibly halted photosynthesis entirely. At the time of the impact(s) an enormous amount of heat would be generated, providing the potential for setting off huge brush fires and the like. Charcoaled horizons have been detected which might represent such events. More recently, shocked quartz crystals – another sign of impact – have been discovered in the same horizon, adding support to the impact idea.

The evidence for a catastrophic event is mounting, but this is not to say that it alone supplies the answer to the end-Cretaceous mass extinctions. Unfortunately, some scientists have somewhat over-enthusiastically grabbed hold of the idea, giving it credit for almost everything. Given the stratigraphic and palaeontological evidence for what had been happening to genera and species during the latter part of the Cretaceous, there is little doubt that an asteroid impact did not simply wipe out the dinosaurs; period! Nevertheless, while it would be quite wrong to assume that the catastrophe was the complete answer to the K/T mass extinctions, it might quite reasonably be argued that an asteroidal impact was the single event which speeded up the final demise of those species known to have declined and then disappeared.

Backtracking for a moment: careful study of the rock record makes it quite clear that the pelagic plankton disappeared in around 100 000 years – very

rapidly on the geological time scale. Other genera appear to have gradually declined, indeed they probably declined in a series of pulses, over a period of around 10–15 million years, in some cases, far longer. The effects of a catastrophic event (an impact) would be expected to occur over a period of less than 1000 years. Therefore a catastrophe event alone does not account for the observed time scale of change.

Are there any other possibilities? Well, there certainly is one process which can cause rapid climatic change, and might explain the isotopic evidence, or some of it. This is an endogenous geological process – massive effusion of magma. The most major volcanic outbursts, such as the ones which emplaced the sequence of basaltic lavas known as the Deccan Traps, or the Columbia River Basalts, are known to send up large amounts of carbon dioxide and dust into the upper atmosphere, to be associated with vigorous plate activity and to be capable of very effectively modifying global climate over a relatively short period. Since it takes but a few hundred thousand years for their destructive effects to run their natural course, then here we have another possible contributor towards biologic extinctions. Again, however, volcanism alone could not satisfy all the facts which have been brought to light.

Research continues into the cause of this mass extinction. Stratigraphers, palaeontologists and geochemists are, frankly, divided in their opinions. Much more work will have to be done before we can possibly say that we have the complete answer to the problem. Nevertheless, whatever the truth turns out to be, it is clear that the notion of uniformitarianism – things have always been the same as they now are – is a more questionable principle than once was believed. It seems that catastrophe has played a vital part in evolution, there having been other mass extinctions during the Cambrian, Ordovician and Permian, to name but three.

21

New oceans for old

The modern oceans

The break-up of Gondwanaland caused dramatic changes in the configuration of land and sea. Before the onset of the Mesozoic plate activity that brought this about, the Pacific and Tethys Oceans were the Earth's principal areas of sea. Of these two, the latter separated Laurasia from Gondwanaland, and has since been virtually eliminated save for two remnants, the Caspian and Black Seas. The Pacific, on the other hand, still remains the largest ocean, although it is slowly decreasing in area due to subduction along its western margin. Although it is floored predominantly by crust that is less than 200 million years old, there has been a Pacific to the west of the Americas for much longer. While it may not be as old as the Moon – which once was believed to have been spawned from its depths – it does date back to the Early Palaeozoic, and possibly may be even older. Naturally the oldest Pacific crust has long since been recycled inside the Earth, but it remains true to cite it as the Earth's oldest remaining ocean.

In contrast, the Atlantic, Indian and Antarctic (or Southern) Oceans are all relatively youthful, growing as Gondwanaland broke up and its components drifted slowly apart. Eventually the modern pattern of land and sea was achieved, ocean growth being made possible by the generation of new oceanic crust

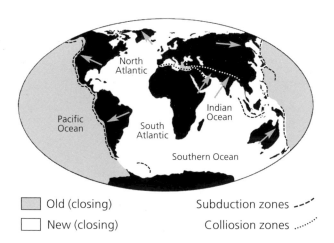

The opening and closing oceans.

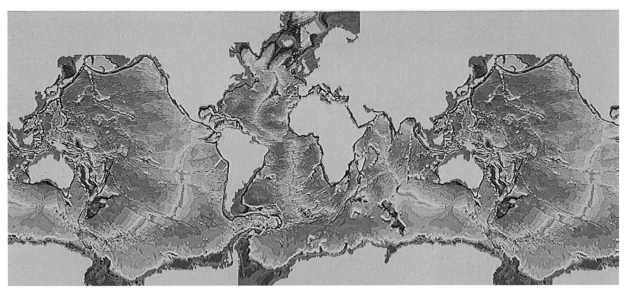

Bathymetric map of the world's oceans as acquired by the SEASAT satellite. Computerized contouring beautifully brings out the main submarine structures, including oceanic ridges and transform faults.

at mid-oceanic ridge systems; the Mid-Atlantic and Carlsberg Ridges being two of these (the latter lying within the Indian Ocean). The newly-generated crust is constantly being added to their following margins as time passes by.

One significant difference between these new opening oceans and the Pacific is their plate tectonic situation; the Pacific is almost entirely encircled by subduction zones, the adjacent continental crust not being coupled to the oceanic crust. This means that although active spreading is occurring along the East Pacific Rise, the oceanic lithosphere at the margins of the ocean can be consumed constantly at the peripheral subduction zones, allowing the total area of oceanic crust slowly to decrease. With younger oceans, such as the Atlantic, things are quite different; here the lithospheric plates are composite, carrying both oceanic and continental crust. Since there

Iceland: recent cinder cone with low basaltic shields in background. [Photo: Bob Toynton]

are no subduction zones associated with the Atlantic, it continues to grow. I have entitled this chapter 'New oceans for old' – because young oceans are expanding at the expense of their predeccessors.

The Mid-Atlantic Ridge

The Mid-Atlantic Ridge runs as a continuous mountain range along the axis of the Atlantic, marking the suture along which the African, North American, South American and Eurasian plates separated. The crest of the ridge is offset in many places by cross-cutting fractures. These include inactive fracture zones and active ones termed *transform faults*. The latter often are seismically active, particularly where they intersect the ridge. The fractures take up the strain imposed on the relatively brittle oceanic crust by the ductile mantle beneath, in a sense enabling the spreading centres to 'bend'. Strangely, the Mid-Atlantic Ridge is at once the oldest and youngest part of the ocean; while it is geographically old, the youngest rocks are being created along it.

The least complex part of the ridge is the central portion, which parallels the curves of the African and American continental margins; to the south of this, it is offset by several fracture zones then, south of the Equator it straightens and eventually turns east around the southern tip of Africa, eventually linking with the Indian Ocean ridge. To the north, however, things get rather more complicated; the ridge passes through the Azores and runs between Newfoundland and Spain. Between Labrador and Ireland it is offset along the Charlie-Gibbs Fracture Zone, then striking northeast to form the Rejkanes Ridge which runs through Iceland and was the first mid-ocean ridge segment at which symmetrical magnetic lineations were discovered and used to confirm the notion of sea-floor spreading.

Iceland is the largest land region made from oceanic crust, and is the site of a negative gravity anomaly, which strongly suggests that mantle material must lie at unusually shallow depths. The island is built from plateau basalts and lavas extruded from central volcanoes, and there are also thousands of dykes, mainly of basalt. Generally speaking, Icelandic lavas get older with increasing distance from the island's axis – away from the active spreading centre. The oldest lavas so far dated by radiometric means yield ages of about 16 million years – very much younger than the lavas of East Greenland and the Faeroes (60 million years).

Iceland itself is a part of the ridge and also is part of a transverse topographic rise which joins the ridge to Greenland and the Faeroes Islands along the Arctic Circle. During Tertiary times a tremendous episode of igneous activity took place here, producing extensive plateau basalts in the Faeroes, East Greenland and Northern Ireland. North of Iceland the ridge continues to the volcanic island of Jan Mayen, then is offset to the east; it passes close to the west coast of Spitzbergen, then, via a series of offset faults, into the eastern part of the Arctic Ocean. There it can be traced beneath the ice cap until apparently it dies out near the Lena delta of the Asian continent.

The Red Sea and Gulf of Aden

The Red Sea and Gulf of Aden are of particular interest in the context of young oceans which are in process of spreading. The axis of the Red Sea is an active spreading centre, the entire region being one of high heat flow, recent continental volcanism and current seismicity. Rifting probably commenced in the Late Tertiary and, for at least the past 10 million years the sea floor has been spreading at the rate of about 0.8 cm per year. Oceanic crust has floored the Red Sea for at least 5 million years. Along the margins of the Red Sea are recent continental volcanic rocks; there are also very thick salt deposits: both are characteristic of

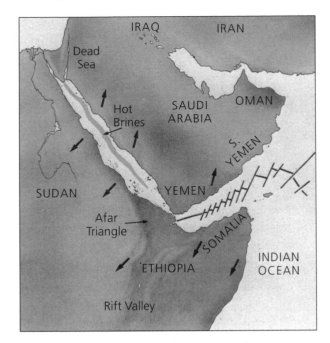

Map of the Red Sea and Gulf of Aden region. Alfred Wegener observed that if the Afar Triangle is considered a recent addition (which it is), the two sides of the continent fit quite snugly together.

The Red Sea itself is an elongated trough 1800 km long. It is remarkably parallel-sided in the north, then widens, and subsequently narrows again to the Straits of Bab El Mandab, at its southern extremity. There are two topographic units: a shelf which is between 100 and 200 m deep and a central trough which generally lies between 600 and 1000 m down, but in one place descends to 2200 m. Beneath the trough the heat flow is very high indeed because young oceanic crust underlies it.

To the north of this region are the mountain ranges of the Taurus and Elburz which run eastward through Afghanistan and into the mighty Hindu Kush. These were formed when the remnants of Gondwanaland (Arabia and Africa) moved northward and collided with Laurasia during the Mesozoic era.

newly-splitting continents. In the Gulf of Aden the oceanic Carlsberg Ridge (which runs across the northern part of the Indian Ocean) is offset by several transform faults. To the west, the rifting continues into the East African Rift system.

View of Trap Series rocks, Manakha, Yemen. Tertiary flood basalts were poured out from fissures in the continental crust as rifting began at the incipient continental margins of Arabia and Africa, when they began to split apart.

This is when the ancient Tethys Ocean was squeezed out of existence.

The oldest rocks found in the region are exposed in Saudi Arabia, on the shoulders of the Rift. These are mainly metamorphic and igneous rocks whose ages range from 1100 to 1300 million years BP. Similar rocks are found on the western side, in Africa. An interesting feature of these old rocks is that they have a distinct north–south structural grain, as if alternate strips of country have previously acted either as topographic 'highs' or 'lows'. At several times in the past, magmas have risen through fractures separating the blocks, emplacing a variety of igneous rocks. The main sets of dykes so produced have been dated at 500 and 300 million years BP. The modern Rift follows the same structural orientation and it seems that the modern Red Sea rifting followed a pre-existing structural line which had been in existence for at least 1000 million years.

The precursor to all of this new ocean activity appears to have been the doming of the region centred on the Red Sea–Gulf of Aden–Ethiopian Rift systems, about 40 million years ago. This became a focus of intense magmatism around 35 million years ago, seeing the outpouring of vast floods of basalt and injection of dykes into the ancient Arabian and African cratons. Now, after splitting has taken place, these lavas lie on opposite shores of the sea. In the Gulf of Aden and Arabian Sea regions the opening of a rift took place somewhat earlier than in the Red Sea, with oceanic crust having existed here for at least 5 million years longer. The spreading axis which runs along here is offset by several transform faults and eventually runs into the Indian Ocean ridge system.

The Indian and Southern Oceans

At once separating yet linking the younger Atlantic and more ancient Pacific Oceans, are the Indian and Southern Oceans. From the abyssal plain rises an oceanic ridge system that delineates an inverted 'Y' and defines the boundaries of three diverging lithospheric plates. The Carlsberg Ridge runs north to south between Peninsular India and the eastern coast of Africa, and is diverted into the Gulf of Aden at its northern end by a series of transform fractures. To the south, the ridge forks, the westerly branch trending southwest at first, then west to run between Africa and Antartica, eventually joining the Mid-Atlantic system at its southern extremity. The easterly branch strikes in a direction roughly eastsoutheast between Australia and Antarctica.

Palaeomagnetic traverses made across the ridge to the south of Australia indicate a spreading rate of about 3.5 cm per year on either side. Using this information and correlating magnetic stripes here with those in the Pacific, it is clear that the ocean separating Australia from Antartcica has spread for only 35 million years.

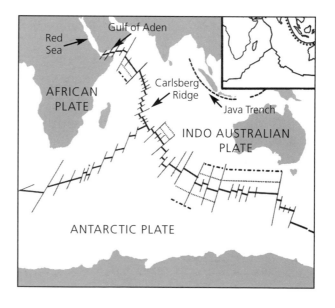

Map of the Indian Ocean, showing oceanic ridges and transform faults. The boundaries of the lithospheric plates are shown in the inset diagram.

Deep-sea cores and seismic traverses together show that very large areas of the ocean bed on either side of the ridge system are clear of sediments; something which suggests that it is very youthful. The thickest sediment accumulations are found in the Southern Ocean, where *oozes* made from siliceous organisms called diatoms, give rise to sequences which may be anything from 100 to 750 m thick. Land-derived debris is most evident on the Indian Ocean floor off the mouths of the great Ganges and Indus rivers, where vast submarine fans have been under construction ever since the Himalayas began to rise. The average rate of accumulation of this river-borne material is of the order of 17 cm per 1000 years. Most of the debris is well stratified and consists of turbidites.

Other substantial sedimentary accumulations are found off the east coast of Africa, where thick sediments have accumulated in the Somali and Madagascar Basins. Madagascar itself is a fragment of continental crust surrounded by ocean floor. The very ancient gneisses which form the island's backbone are overlain by Late Palaeozoic and Mesozoic rocks which collected in fault-defined basins. The Seychelles also have continental characteristics, being built largely from Early Palaeozoic granites, and are believed to have formed part of the continent of Gondwanaland before being rifted away during continental fragmentation.

On the northeast side of the Indian Ocean is an island arc system. This emerges as the island arcs of Indonesia (Sumatra and Java being the more westerly islands). To the south and west of these are deep trenches and sediment-laden troughs that are associated with northward-dipping Benioff Zones. It is along this line that the northward-moving Indo-Australian Plate is in contact with the Eurasian plate.

The extreme distortion and crumpling of the island arc system is a result of the tectonic interaction between the two plates.

The Atlantic's margins

Earlier I described how, prior to the fragmentation of Gondwanaland, doming and fracturing of the continental crust preceded a phase of magmatism, giving rise to vast tracts of plateau basalt lavas and intrusive igneous rocks. As the crust continued to stretch, faulting and cooling of the crust brought about the formation of marginal basins, these now residing along the margins of the modern continents. Similar events are recognizable associated with the Atlantic Ocean.

In the early stages of its development, when the youthful Atlantic was little more than a shallow and very restricted trough, sea-water would have flooded into it, only to be evaporated due to the restricted circulation and shallow depth of water. Thus the first marginal deposits were Triassic evaporites. As the ocean widened, marine shelf deposition took over

Nimbus-7 coastal zone colour scanner image of the Western Approaches to the British Isles. The yellow and brown hues denote high levels of sediment in the surface waters of the continental margin. The blue colours are probably due to plankton 'bloom'.

and, as the thickness of the sediment pile increased, the amount of depression of the continental margins increased and the basins sagged more and more. Where large river systems flushed huge amounts of land-derived debris out onto the ocean floor, as for instance was the case with the Amazon and Niger rivers, massive deltas built out, aiding and abetting this process.

Deep boreholes have been sunk into the crust along the Atlantic seabord of North America, revealing how sedimentation began as soon as continental separation had started. Basin sequences as thick as 3 km are quite normal. The continental shelf regions are often found to be incised by deep canyons, into which sediment is flushed by turbity currents from the unstable shelf regions on either side of their head regions. These were first documented along the coast of Newfoundland, where turbidity current flow frequently sundered the submarine telephone cables.

Along its eastern margin the modern Atlantic has a number of large basins receiving normal marine sediments. The North Sea basin is in reality a kind of 'failed ocean', and is somewhat atypical. However, to give some idea of how much sediment can accumulate in a fairly young basin, off the coast of Holland, the Mesozoic and more recent infill is 6 km in thickness and is pierced by salt domes which act as very efficient traps for the hydrocarbons which have collected here. Other basins lie off the Western Approaches, and there are also the Aquitaine and Lusitanian Basins further south.

The western margin of the Atlantic, however, is complicated by the presence of short island arcs at two isolated locations along the otherwise passive continent–ocean boundary. The Lesser Antilles form an active island arc, separating the Atlantic from the Caribbean (a separate small ocean basin). In the far south, another active zone – the Scotia Arc – links the active orogenic belts of the Andes and Antarctic Pensinula. These two small regions of subduction are exceptional, however, and the Atlantic margins fundamentally are of the passive kind.

Finally, one feature which very effectively spoils the otherwise good 'fit' between America and Eurasia, is the Iberian Pensinula. All attempts to reconfigure the continents prior to continental drift have had a problem with this part of the world. Its position, and the existence of the Bay of Biscay, are explained by the palaeomegnatic data, which reveal that Eocene lavas in Spain have a rather different magnetic orientation to similar rocks from elsewhere on the continent of Europe. However, by rotating Iberia about 40° anticlockwise, the orientations can be brought into line. This makes it clear that the peninsula rotated 40° anticlockwise – away from France – during the opening of the Atlantic.

22

The Pacific Ocean

An ancient ocean

Despite the fact that there is oceanic crust older than about 200 million years beneath the Pacific, the feature itself is much older than this, its history spanning the whole of the Phanerozoic and probably even before that. Among the most fundamental differences between it and, say, the Atlantic is, as we have seen, its virtual encirclement by active subduction zones. These have acted out a major rôle in Earth's history, as they have made it possible for a balance to be maintained between the creation of new crust and the destruction of a sufficient amount of older crust to maintain the correct balance.

Whereas in most oceans the spreading centres are symmetrically positioned, this is not the case with the Pacific. The main ridge enters the ocean from a point between the southern tip of New Zealand and Antarctica, then, at about longitude 120° W, turns northward and runs approximately south to north towards California, as the East Pacific Rise. This great submarine mountain range is among the most closely studied parts of the ocean floors, and has been criss-crossed by more oceanographic vessels than most, using the most up-to-date geophysical devices, and inspected at close quarters by scientists in submersibles.

An artist's impression of lava domes along a mid-oceanic ridge.

high and bisected by a rift; on either side it slopes down to the abyssal plain. The comparatively broad nature of the rise is likely to be a function of the higher rate of spreading experienced here: magnetic striping indicates different rates from place to place, with the average being around 4.5 cm per year. This is substantially higher than in the Atlantic. Using this rate in calculations shows that a mere 40 million years only would be required to generate the entire floor area of the eastern Pacific.

Along the northern section of the rise – west of central America – the structure is rather complex: numerous transform faults offset its crest, these being spaced between 200 and 300 km apart. Between the principal fractures are minor ones which may be separated by as little as 10 km. Close study of these has shown that near major transform faults the rise is quite deep below sea level, while it rises to much shallower depths roughly mid-way between them. In other words, between each pair of major fractures the structure is roughly that of a broad lava dome; smaller domes – more like large blisters – occur between the smaller faults, and it appears that each of these is a miniature spreading centre. What we seem to have is, rather than just a few major spreading centres, a plethora of rather small ones. This was hitherto unsuspected.

Seismic studies indicate the rise to have deep roots going down to at least 200 km, where the Low Velocity Zone is encountered. The Low Velocity Zone, where seismic waves travel at their lowest velocity, is normally found at half this depth. Beneath are large magma chambers of gabbroic composition, the coarsely-crystallized plutonic rocks sometimes reaching the surface as blocks within basaltic flows. In an effort to understand more about the structure, a deep borehole was sunk into the ocean floor in the vicinity of the Costa Rica Rift during 1981; this particular fault crosses the rise between South America and the volcanically-active Galapagos Islands.

As is the case with the Mid-Atlantic Ridge, the rise is offset by numerous transform fractures, several of which connect the rise with an active spreading centre in the Gulf of California. From then on, apart from minor spreading centres, there is no active sea-floor spreading offshore of the western USA. The distinct asymmetry of the ocean about the East Pacific Rise is due to the fact that the oceanic crust to the east has been overriden by the North American continent, the original spreading centre having now gone underneath the continental crust. Further south, the oceanic plate to the east of the rise is being actively subducted beneath South America. Such a situation is not found with the Atlantic, Indian, Southern or Arctic Oceans.

Oceanic ridges and rises

In recent years we have learned a lot about the Pacific spreading centres, in large part due to the investment in research by commerical enterprises interested in the rich sulphide deposits known to exist there. Furthermore, there have been numerous dives by the French submersible, *Cyana*, and others, which have given us a much clearer idea of how new oceanic crust is being formed.

Profiling across the East Pacific Rise has shown this to be a much broader structure than the Mid-Atlantic Ridge. Atop the crest is a 30-km-wide ridge, 500 m

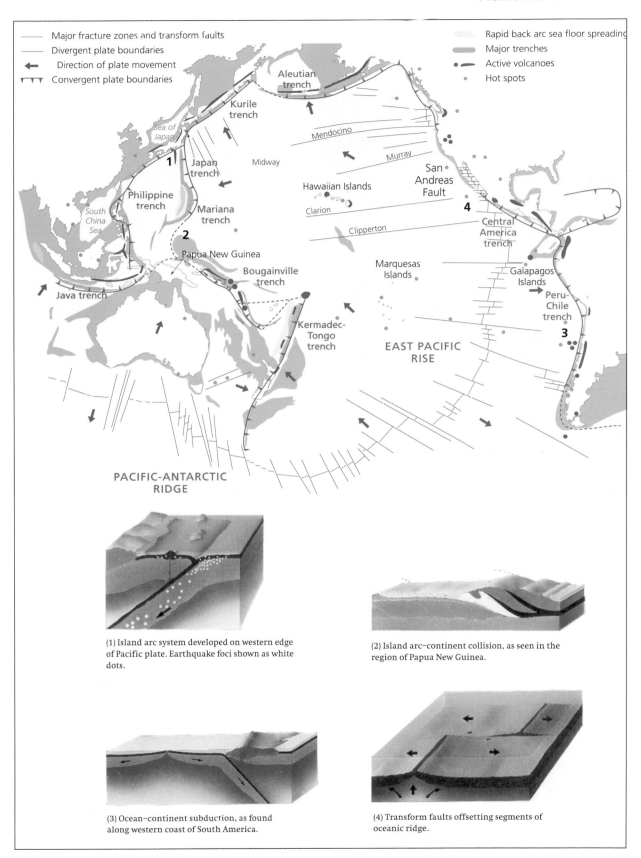

(1) Island arc system developed on western edge of Pacific plate. Earthquake foci shown as white dots.

(2) Island arc–continent collision, as seen in the region of Papua New Guinea.

(3) Ocean–continent subduction, as found along western coast of South America.

(4) Transform faults offsetting segments of oceanic ridge.

Chart of the main features of the Pacific Ocean, with descriptive cross-sections.

The borehole, known as 504-B, penetrated the sea-floor to a depth of 1076 m and, although this may not sound particularly impressive, was a major achievement in view of the very difficult conditions encountered in true deep-sea drilling. The samples have subsequently been radiometrically-dated and the crust shown to be 6 million years old. The core also revealed that in this neighbourhood the ocean floor is mantled by 275 m of recent sediments which are underlain by 575 m composed of pillow lavas and brecciated volcanic rocks of basaltic composition. Similar rocks extend below this level and down to 780 m, within which zone they are cut by hundreds of basaltic dykes. Below this point the pillow lavas give way to massive basalts and even more abundant dykes. This may not be representative of the structure of the oceanic crust beneath the entire Pacific Ocean, but is likely to approximate it over large areas.

Islands arcs of the Pacific – Japan

Arcuate chains of islands festoon the northern and western margins of the Pacific, extending from the Aleutians through Japan to Ryakyu. They then extend southwards through the Phillipines, turning south-eastwards towards New Caledonia to New Zealand. On the boundary between the southwest Pacific and the Indian Oceans lie the complex arcs of Indonesia. The arc-like form is not their only common feature, they are seismically and volcanically active and are associated with deep trenches that lie above inclined Benioff Zones. The latter are the planes along which oceanic crust currently is being subducted and recycled inside the Earth; they are inclined at around 35° to the horizontal. Moving towards the active volcanic arc from the trench, the depths of earthquake foci

gradually increase. Island arcs are a feature of convergent plate boundaries.

Of all the Earth's island arcs, the Japanese Islands have been most intensively studied, and the findings used to understand other similar structures. A first glance suggests that the islands might best be considered to form four different arcs, but when submarine topography is analysed, it is found that there are, in fact, only two. One – the East Japan Arc – joins the Kurile, northeast Honshu and Izu–Bonin–Marianas group of islands, the other – the West Japan Arc – links the Ryukyus, Kyushu, Shikoku and western Honshu groups. On the ocean-facing flanks of both arcs are found the deepest trenches. Asia sits well back from the active arcs, and between the island groups and the continental margins are well-devel-

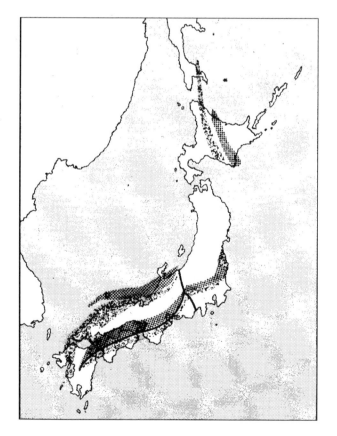

The metamorphic belts of Japan. The Median Tectonic Line and Fossa Magna are shown as heavy lines. High-temperature/low-temperature belts (dotted) and low-temperature/high-temperature belts (stippled) are also shown.

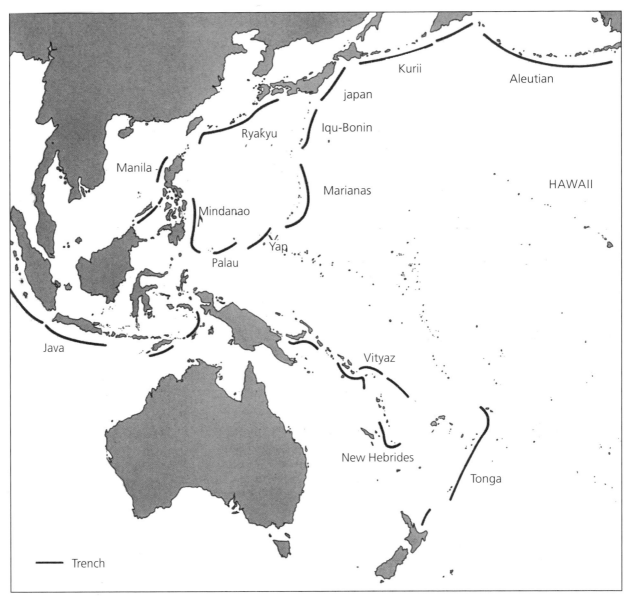

Map showing island arcs and trenches of the Western Pacific.

oped marginal seas, such as the Sea of Japan.

There are marked differences between northeast and southwest Japan. The former shows features that are typical of active arcs, i.e. volcanic activity and the presence of a deep trench; the latter is volcanically quiescent. In fact the volcanic belt of northeast Japan lies along the landward side of a major structural discontinuity called the Fossa Magma, which it then turns into before extending southward into the Izu–Bonin–Marianas arc. Southwest Japan, on the other hand, has a 'grain' that trends roughly parallel to the island's axis and includes metamorphic belts which occur in pairs.

Slightly to the landward side of the deepest trenches there are negative gravity anomalies which indicate a mass deficit. This implies that rocks of low density are present beneath the arc. This low-density material must be the oceanic crust of the Pacific floor which is descending into, and displacing, the denser rocks of the underlying mantle.

Island arcs of the Pacific – Indonesia

The Indonesian archipelago is 6400 km long and has

a total land area of 5 million km², while its sea area is three times this. It includes amongst its 13677 islands, three of the ten largest islands on Earth: Papua New Guinea, Borneo and Sumatra. Most Indonesians live within sight of a volcano, three earthquakes are recorded within its region every day, and there are 400 volcanoes of which between 70 and 80 are active at the present time. It is one of the Earth's most active (and dangerous) regions. Two of the Earth's largest ever cataclysms – the eruptions of Krakatoa and Tambora – were felt here. It is unusual and extremely complex in its geology because it comprises not one but two active arcs – the Sunda and Banda Arcs. Only the simplest explanation will be attempted here!

This series of island arcs has formed in response to the movement of three large plates: the Eurasian, Pacific and Indo-Australian, plus a smaller Phillipine Sea plate. Currently the Indian plate is sliding northward beneath Sumatra and Java at a velocity of about 6 cm per year. If the relative motions of these three plates continues as now, say for a further 30–50 mil-lion years, the island systems of Indonesia and the Phillipines will likely be squeezed between a colliding Eurasia and Australia such that the newly-enlarged Eurasia will contain another broad and intricate selvage of mashed-together island arcs and continental fragments (for that is what Papua New Guinea is) set between internally-stable subcontinental blocks. The features of this active subduction system are: trench–outer-arc ridge–outer-arc basin–volcanic arc–foreland basin.

All of the islands of the chain lie on the broad Sunda Shelf which extends down from the mainland of Southeast Asia. The intervening Java Sea is quite shallow (less than 150 m deep) and was, during the last ice age, a huge subcontinent. It is roughly 350 km wide. Papua New Guinea and the neighbouring islands of the Arafura Sea are connected in a similar way to the continental shelf of Australia; indeed, Papua New Guinea was torn from that continent in the distant past. In between these two shallow continental shelves sit the islands of Sulawesi, Maluka and Nusa Tengarra which rise from a deep trough created by faults associated with plate convergence.

The outer edge of the Java–Sumatra Trench is 1 km deep, plunging to 4.5 km off North Sumatra and 6 km off Java. The abyssal sediments become thicker towards the shallower regions, where the oceanic basement also thickens. Most of the land-derived sediments from Java are trapped close to the island in an *outer-arc basin*. If they actually reach the trench itself, they quickly become incorporated into the unstable prism of sedimentary material called the *mélange wedge*. East of Java the ocean floor sediments are again thick, having been washed here from Australia; these become involved in the subduction process, and water depths of the ocean floor and trench again become shallow as the sediment prism becomes thicker.

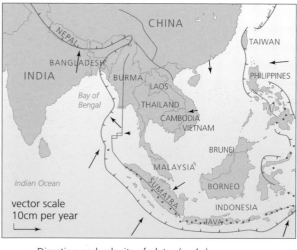

→ Direction and velocity of plates (cm/yr)

Trace of subduction zone (sea) or thrust belt (land)

• Active volcano

0   1000   2000km

General structural map of the Indonesian island arc system.

Aerial view into crater of Gunung Bromo, Java, a typical island arc volcano.

The ominous cone of the andesitic subduction volcano, Gunung Merapi, steaming vigorously above the densely-populated city of Yogyakarta, Java. It has been periodically responsible for much devastation and loss of life in the vicinity, largely due to the generation of ash, glowing avalanches and lahars.

The volcanic caldera of Gunung Batur, Bali, showing stratification in the caldera walls and a part of the floor with recent lava flows (dark). Viewed from the recently-active cone of Gunung Batur, located at the centre of this huge depression. This volcano was last active during 1999.

The continental slope rises from Java and peaks in an *outer-arc ridge*. The ridge is wholly submarine off Java, Sumatra and the other islands along the volcanic arc; its crest lies between 1 and 3 km down. Towards the northwest, however, it rises and is only 1 km below sea level along most of Sumatra. Both the continental slope and this ridge are underlain by a landward-thickening wedge of mélange and imbricated sediments.

Much of the sediment washed off Java and Sumatra during the fierce tropical storms accumulates in an *outer-arc basin*. This is separated from the trench by the outer-arc ridge just described. The smoothness of its profile and its floor indicate that turbidites flow along it longitudinally. It is a broad basin with water depths of between 2 and 3 km off Java but only 200–300 m off most of Sumatra. Thick sediments lie beneath most of it. These are little deformed in the centre but become progressively more deformed towards the ocean, sometimes becoming actually overturned. The flat-lying sedi-

ment accumulations along the basin's axis have been shown to be 6 km thick!

The thin wedge of deformed sediments beneath the landward slope of the Java trench is imbricated at a steeper dip than the dip of the subducting plate. It seems that the sediments are being deformed by some oblique shearing process in response to the couple represented by the landward subduction of the plate beneath, and the trenchward gravitational movement of the deforming wedge above. The wedge thus grows by the scraping-off of parcels of deformed sediments at its front.

Above the subducting slab of oceanic crust and its veneer of marine sediments rises the active *volcanic arc*. This is characterized by often violently-explosive basalt–andesite–dacite–rhyolite volcanism. What appears to be happening beneath the arc is that the oceanic crust is dehydrated, water being driven from the descending slab into the mantle wedge above, taking silica and potash with it. The resulting magmas that are formed by the heat generated, become enriched in these elements, moving them away from a mantle chemistry. If a high degree of partial melting occurs, then basalts are generated; if only a low degree takes place, then andesites and the more silicic magmas are produced. Either way, some extremely dangerous volcanoes exist here: Krakatoa, Merapi, Agung, Ijen, Rinjani and Tambora to name just a few. Their eruptions may involve not just the explosive eruption of ash clouds, but also glowing avalanches and mudflows; all have the potential to cause great devastation and sometimes loss of life. Some of their eruptions are described in Chapter 26.

The Papua New Guinea story

This huge island lies at the eastern end of the Indonesian Archipelago and closely north of Australia. It has a mountainous backbone with permanent glaciers and an extremely complex geological history. There are several active volcanoes and seismicity. To the southwest of the Papuan Pensinula is a submarine plateau which lies at a depth of around 2 km and appears to be a sediment-coated fragment of continental crust.

Southern Papua New Guinea is in reality the northern part of the Australian craton, separated from it only by a shallow sea. Southwest Papua New Guinea was a craton during the Palaeozoic, while the orogenic belts of eastern Australia can be traced northwards into central Papua New Guinea and once went even further, along the Pacific side of a continental block now comprising Sumatra and the Malaysian peninsula. This was rifted away, however, during the Mesozoic.

During the Late Mesozoic and Early Tertiary period, the northern edge of the Australian–Papua New Guinea landmass was a stable continental shelf. Australia and Papua New Guinea were separated from Antarctica during Eocene times, and since then have been drifting northwards. Until Miocene times the northern continental margin was a stable shelf area; however, thereafter an island arc which had been migrating southward during Cretaceous and early Tertiary time, collided with the continent, generating mélange wedges and thrusting slabs of ophiolite onto the crust between the continent and the accreted island arc. The continent was still moving northwards at this time but after the late Cenozoic the direction of subduction reversed, i.e. it was southward beneath the enlarged continental landmass. The eastern part of the island arc, where not stopped by continental crust, continued to drift southward and now forms the Papuan Pensinula (the southeastern tip of the island).

The history of the Papuan Peninsula is interesting. It moved southwestward during late Tertiary times, overriding the oceanic crust of the Coral Sea and Gulf of Papua. This was a part of the story of the southward-moving island arc described above. In other words, before that time (Miocene), the continent of Papua New Guinea only existed to the west of what is now longitude 145° 20′ E. The strata deposited on both the continental slope and shelf were then shoved southwards onto the continental margin by the colliding arc, and this was the point at which subduction flipped to the southward direction. Of course, to the east of the continent, the arc felt no resistance to its continued southward drift, and it was able to sweep before it the sediments on the floor of the Coral Sea, which was subducting at the time.

The east Pacific margin

This margin of the Pacific is very different from the others. Here the Pacific has lost ground to the westward-moving plates that carry the Americas and in consequence the character of subduction is rather different. In the Gulf of California, and northwards from there, the boundary between the Pacific and North American plates is offset by numerous trans-

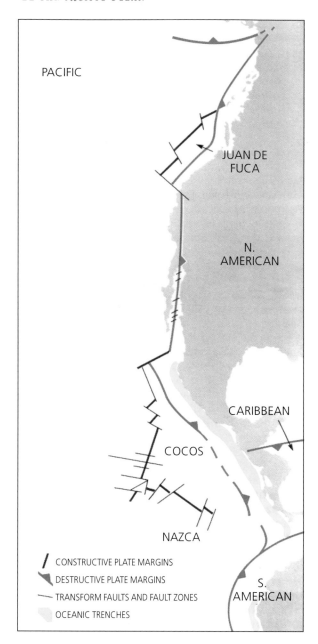

CONSTRUCTIVE PLATE MARGINS
DESTRUCTIVE PLATE MARGINS
TRANSFORM FAULTS AND FAULT ZONES
OCEANIC TRENCHES

The convergence zone between the Pacific, Juan de Fuca and North American plates.

Plate boundaries to the west of Central America, showing the junction between the Cocos Plate (centre), Pacific Plate (left) and Nazca Plate (bottom right). Spreading axes shown in red, convergent plate margins by toothed lines.

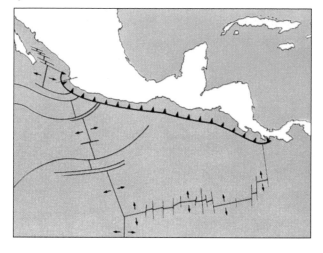

form faults. Earthquake foci typically are quite shallow (10 km) and are concentrated both along the boundary and the transverse fractures which offset it. It is here that we have the infamous San Andreas Fault.

Although the individual motions along the San Andreas Fault (and others like it) are quite small, their cumulative effect over a period of a million years or so, is considerable. Even during the famous 1906 San Francisco earthquake the horizontal displacement was a mere 5 m; during the earlier quake in the Owens Valley, the vertical movement was only 4 m. However, if similar movements are repeated many times over, substantial vertical and horizontal displacements of the crust could be accomplished in a million years.

During Jurassic and Tertiary times, the North American continent overrode the eastern floor of the Pacific, and there are currently neither trench nor deep-focus earthquakes along this stretch of the ocean's margin. Today, one plate – which comprises the majority of the North Pacific Ocean together with a slice of western California – is in contact with the North American plate – which includes the remainder of North America and the western portion of the Atlantic. The San Andreas Fault separates the two along much of the western coast of North America. Because the Pacific plate is moving inexorably northwestwards with respect to North America, and towards the Aleutians trench, so California experiences frequent earthquakes in response to these subcrustal movements.

The East Pacific (Farallon) Plate – which now has been largely subducted beneath North America – nevertheless still leaves two remnants behind: one lies south of the San Andreas Fault and is called the Cocos

Plate, the other is a smaller fragment which intercepts the American coastline between northern California and Vancouver; this is called the Juan de Fuca Plate. Northeasterly or easterly subduction is still ongoing along those stretches of the western coastline where these remnants abut against it.

As the East Pacific Rise gradually was overrun by the North American Plate, so an unstable triple junction developed where the three plates met (i.e. the existing Pacific Plate, the largely overrun East Pacific Plate – now represented by the Cocos remnant – and the North American Plate). This triple junction gradually migrated along the continental margin as more and more of the Pacific Plate came into contact with

North America. An active triple junction exists further south, where the Pacific, Cocos and Nazca Plates meet.

The situation even further south is quite different. Here there are deep trenches off the western coasts of both Central and South America. Furthermore, Benioff Zones dip steeply under South America. Branches of the East Pacific Rise trend eastwards towards Central America and also towards southern Chile; these define the smaller Nazca Plate which currently is slipping beneath the South American plate. The story of the Andean cordillera, formed in direct response to these plate movements, is described in the next chapter.

23

The cordilleran chains

Mountains of the western Americas

In Chapter 14 I discussed the geological development of the North and South American cratons during Pre-Cambrian and Palaeozoic times. Thus far I have not, however, looked in more detail at the western margins of these continents. This omission must now be rectified.

The events which fashioned the mighty ranges of the western Americas were in many ways similar to those which generated both the Caledonides and the Appalachians, but whereas volcanic and tectonic activity is now minimal in these regions, there is considerable unrest in the west – witness the eruptions of Mount St Helens and El Chichon. This tells us that ongoing plate activity is a feature of this region at the present time.

The high mountain chain that runs close to the western perimeter of both continents from Alaska to Mexico, and from Guatemala to the southern tip of Chile, has a length of almost 18 000 km. It is an immense elevated belt which has remained geologically active throughout the Phanerozoic. The two continents have not always occupied the same position relative to one another during that period, the modern disposition being a result of relatively recent drifting about following the break-up of Pangaea. Indeed, during Palaeozoic and probably in earlier times, they formed a part of two different supercontinents. About 200 million years ago, the modern Atlantic began to open up, and both North and South America moved westward on the leading edges of lithospheric plates that grew from the Mid-Atlantic spreading centres. It is likely that both continents have been overriding the eastern Pacific Plate since the Palaeozoic, although no collision with other major continental plates has ever occurred. The present mountains are the latest in a succession of chains which have occupied this site, their great elevation being due to recent earth movements.

The Northern Cordillera

The growth of the Western Cordilleras of North America has largely been the result of motions between the Pacific, Juan de Fuca (Farallon) and Cocos Plates. The details, which have been worked out from magnetic anomaly patterns, from radiometric dating and from geochemistry, are still a topic of much debate, but the general picture has become clearer in recent years. Thus, prior to Tertiary times the Farallon Plate had been moving northeastwards, and the Pacific Plate northwestwards. As time passed, more and more of the Pacific Plate came into contact with the advancing North American Plate, such that, about 30 million years ago, an entirely new tectonic situation arose. The North American margin then started slipping sideways against the Farallon Plate – which was by now almost completely subducted – and the Pacific plate. This lateral motion not only distorted the North American Plate but was also capable of displacing it laterally over distances measurable in thousands of kilometres.

Prior to the late Cretaceous (about 80 million years BP), large-scale volcanicity was confined to California and Nevada; subsequently it migrated eastwards until, between 70 and 65 million years ago, it had moved into the central regions of northern Nevada and the Rockies. Then, at the beginning of Tertiary times, the mountains of Colorado were elevated during the Laramide orogeny. The 2-km vertical uplift of the Colorado Plateau, which was accomplished without distortion of its essentially horizontal sedimentary strata, was followed by dissection to give dramatic landscape features such as the Grand Canyon. The central parts of the Rockies were by then rising and, during the Oligocene (35 million years BP), great outpourings of andesitic lavas covered large areas of southwestern Colorado. Then came the major change in the plate movements described above, whereupon large numbers of fissures opened in the brittle continental crust, the Basin-and-Range province being stretched by a factor of two. The extensive Columbia River Basalts were extruded at this time. Finally, western central North America was uplifted about 2 km over an area of 335 000 km^2, to give the modern Rocky Mountains.

Calculations suggest that at least a 700-km-wide strip of oceanic crust must have disappeared down the great subduction trenches of the eastern Pacific, as the Farallon plate was consumed. Many of the vast granite batholiths were produced during this phase of subduction and subsequent melting of crustal

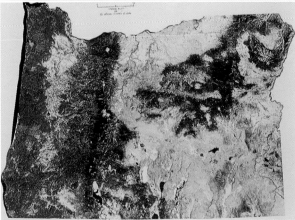

Landsat mosaic of a part of Oregon and Washington states, as prepared by General Electric Photographic Laboratory, Maryland. It shows the western seabord of North America, from the Fraser River delta (top) to just south of Klamath Lake. The Columbia River runs along the state boundary. Very prominent are the western cordilleras, from Mount Baker in the north, through Mount St Helens (north of Columbia River), to Crater Lake (an old volcanic caldera). The volcanic chain has arisen above the contact between the Juan de Fuca and North American Plates.

Map of the general geology of the Western Cordilleras of North America.

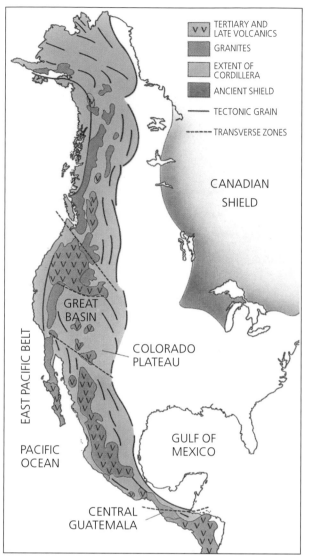

materials. However, while the coast ranges and sierras may be due to subduction, the Rockies clearly were not. These mountains are sited far inland, remote from the line along which oceanic subduction is presumed to have occurred.

Geochemical data from a range of volcanic rocks of this region show that the plate from which the volcanic rocks of the coast range rose was 100 km deep when they were generated. Further east, under the High Sierras, it was twice this depth and, further inland again – beneath the borders of Utah and Nevada – three times as deep. This conforms to the expected pattern of a descending oceanic slab moving eastwards. Surprisingly, however, further inland the lavas apparently were generated from an oceanic plate at much shallower depth, around 200 km.

The implication of this pattern is that there may have been not one, but two subducted plates involved. One possible explanation is that subduction of the ancient Farallon Plate more-or-less ceased when its leading edge had descended to a depth of around 700 km inside the Earth, whereupon it became embedded

The snow-clad summit of Mount Hood, Washington State. This is one of the cascade volcanoes which sits above the subduction zone that underlies this coast of North America. [Photo: Chris Cooper]

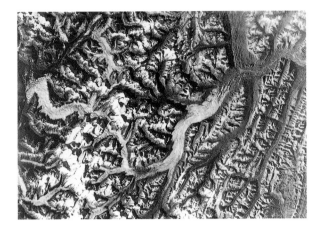

Landsat view of a part of the Canadian Rockies, showing ice-covered ridges and peaks, narrow glaciers and a drainage system which, particularly in the west, is governed by underlying structural control (fold axes and faults). [Photo: courtesy of NASA]

in unyielding mantle rocks. It then broke off beneath North America, so that its shallower end lay along the line of what is now the Wasatch range, a zone of continued instability. Subsequently, subduction started elsewhere, and the uplift of the Rockies and the volcanicity that accompanied this – if the scenario is valid – was due to the heating up and expansion of the buried oceanic plate which lay beneath this region, in response to the elevated temperatures prevailing in the surrounding mantle.

Finally, when the subduction of the Farallon Plate beneath North America had ceased, generation of the andesitic lavas typically produced along converging plates was halted. This effect migrated progressively southwards, as more and more of the

Pacific Plate came into contact with North America. Gradually, as the North American Plate felt less and less pressure from the Farallon Plate, the crust became stretched and voluminous basaltic eruptions took place. These slowly migrated towards the northwest, producing in their path the Columbia River Basalts.

The Southern Cordilleras

Linking the Americas is the archipelago of Central America which resides on the western margin of the North American Plate. Off its western coast are deep trenches that are associated with subduction of the Cocos Plate beneath the archipelago. The present-day topographic expression of the cordilleras of South America is the Andean chain, one of the most magnificent of the Earth's fold ranges. The Andes themselves were raised up quite recently, but it should not be forgotten that orogenic disturbances affected this region as far back as Palaeozoic times. Deep fractures

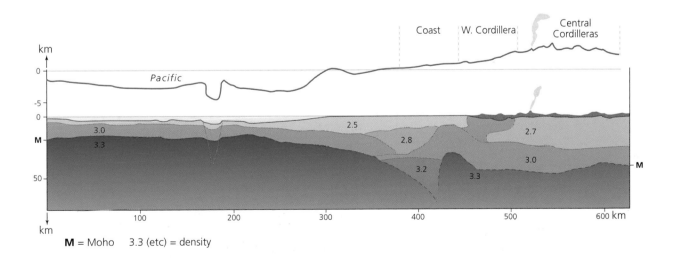

M = Moho   3.3 (etc) = density

Cross-section through the Andean margin of South America.

Space Shuttle image of a part of the Andean chain, showing how narrow it is over much of its length. Here, near Coquimbo in Chile, the coastal plain is clearly seen. The highest part of the range is a mere 140 km wide, the highest peaks rising to 6300 metres. Beneath this section of the Andes the Benioff Zone dips at a low angle, with the result that volcanism is absent. [Photo: courtesy of NASA]

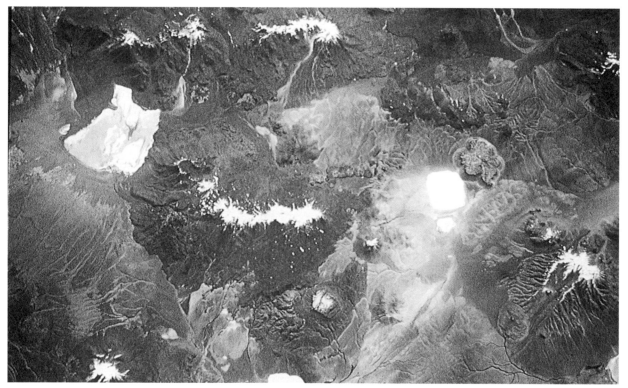

Space Shuttle image of dacite domes and ignimbrite sheets in the central Andean province, Sud Lipez, Bolivia. This is an actively volcanic zone and is extremely remote. Towards the top of the image is a resurgent volcanic caldera traversed by a graben fault (snow-capped and southwest of a playa lake); this is about 4 million years old. A large dacite dome lies close to another playa lake. Pinkish areas are composed of ignimbrite sheets, while several andesite stratovolcanoes (dark) have built up above the ignimbrites. [Photo: courtesy of NASA]

cut the range and are presumed to be graben which extend down to mantle depths. The Andes are rich in ore deposits, most of the valuable ores of copper, silver and gold having been associated with solutions rising from magma sources at depth, these impregnating the higher rocks with their valuable material.

The South American Plate, which has South America on its leading edge, is currently advancing westwards and overriding the oceanic Nazca Plate to the west. The latter is virtually stationary with respect to the Earth's deep interior. Along the plate boundary there are deep trenches which run two thirds of the way down the continent, while to the east, high cordilleras run parallel to the Chilean coastline. A Benioff Zone slopes at 25° eastwards under the continent; this is the site of frequent and often strong earthquakes whose depth increases inland. In some regions of the chain, almost daily earthquakes are the norm; it is a violent region indeed! Approximately 300 km east of the trench lines the cordilleras begin to rise along the western continental margin. Active volcanicity is a feature of the Andean ranges, this being generated because the subducted oceanic slab has attained sufficient depth along this belt for melting to occur. The range is widest in its central sections, where a width of 900 km is attained.

Calculations suggest that the Nazca Plate may be descending at the extraordinary rate of 10 cm per year! If this is the case, then thousands of kilometres of oceanic crust must have been subducted beneath South America during the current phase of plate movement. Particularly confusing in this context is the fact that relatively close to the coast of southern Peru, some 300 km west of what geologists believe to have been an earlier margin of the continent, are continental deposits which are at least 400 million years old (early Devonian). Some scientists have suggested that this strip of coast may be the remains of a very small continent that originally lay off the shore of

South America which was crushed against it as more and more of the oceanic plate was driven eastwards.

In more central regions of the Andes, particularly in Bolivia and Peru, two ranges run parallel to the continental margin. The rocks of the coastal range – or Eastern Cordilleras – formed in the Palaeozoic, in part from marine sediments that had been driven eastwards away from the Pacific. The folding and metamorphism occurred at the close of Palaeozoic times. The Western Cordilleras, on the other hand, are of Mesozoic and Tertiary age and are composed of two belts: the more westerly being built from trench sediments, while the more easterly is made from shallow water marine debris. All of the rocks are highly deformed. This folding occurred in Mesozoic and Cenozoic times.

A lower range – the Coastal Cordillera – runs adjacent to the coastline throughout most of the central and southern part of the chain. This narrow range is an older structural element which contrasts strongly with the higher, younger cordilleras of the interior. The most important folding to affect this range occurred towards the close of Palaeozoic times.

During the Late Mesozoic orogeny, enormous batholiths of granitic rocks rose into the mountain belt and widespread volcanism began. This has continued until the present time. Many of the eruptions are violently explosive, and Mesozoic volcanic sequences are upward of 20 km thick in places. Often these are interbedded with thick terrestrial sediments. Cotopaxi, situated in Ecuador, is the tallest active volcano on land, having an altitude of 5943 m. The last major phase of elevation – which gave rise to the modern aspect of the Andes – appears to date from the late Neogene (10 million years BP). It did not involve any further deformation, but seismic and volcanic activity continues to this day.

The southern part of the cordilleran chain ends in Cape Horn, yet the mountains of the Antarctic Peninsula have a very similar trend and structure.

The similarity is so close that most geologists agree that they were once united. Today they are linked across the South Atlantic by an arcuate chain of volcanically-active islands called the Scotia Arc. To this day geologists are unsure of exactly how the arc was deformed into such an enormous loop, but we can be sure that it has something to do with the tremendous forces unleashed when Gondwanaland broke up.

Antarctica

Antarctica is the Earth's highest continent, more than half of it being over 2000 m above sea level. It is also the coldest spot on Earth and holds two thirds of the Earth's fresh water in its ice cap, thought to be at least 3 km thick. The solid rocks of the Antarctic craton – once a part of Gondwanaland – are exposed in only a very few places, where they show the effects of severe glacial erosion. In recent years considerable geophysical and geological exploration has led to a better understanding of the geology of this remote region. Much of this began during 1956, during International Geophysical Year (IGY).

The continent is roughly circular and 4500 km in diameter. It is divisible into two distinct regions by the Transantarctic Mountains, which stretch from Victoria Land on the Pacific coast, to Queen Maud Land on the Atlantic, a distance of around 2200 km. East Antarctica lies east of the dividing range and is a high plateau that occupies roughly two-thirds of the land area, and rises to a height of 3355 m. This is the ancient Pre-Cambrian shield, its crystalline rocks dating from well before 600 million years BP. West Antarctica comprises Marie Byrd Land and the long Antarctic Peninsula which includes the highest mountain in the two more southerly continents, Mount Vinson, which rises to 5140 m above sea level. The younger fold mountains of this region are considered a continuation of the young fold mountains of the South American Andes.

The ancient shield of East Antarctica is a complex of folded Pre-Cambrian and Cambrian sedimentary and igneous rocks which were refolded during the early Palaeozoic and metamorphosed. They were then intruded by large granitic plutons, prior to peneplanation. The schist and gneiss sequence is very thick, and the metamorphism often to a high grade. Radiometric ages of around 1540 million years BP are yielded by several of the rocks found here.

The Transantarctic Mountains are also composed of gneisses, schists and marbles, and yield Pre-Cambrian–Cambrian ages. Within the sequence are many greywackes which have undergone high-grade metamorphism. These basement rocks are overlain unconformably by flat-lying Palaeozoic–Mesozoic sedimentary rocks. Similar rocks overlie the basement in East Antarctica too, where they attain a thickness of 1500 m. Many of the Palaeozoic sediments are continental in character, including arkosic sandstones and shallow-water quartz-sandstones. One of the most interesting finds in Antarctica was the coal deposits of Palaeozoic age. They tells us that Antarctica was not always at this inhospitable latitude, there having been tropical coal swamps there at the time that it was still a part of the supercontinent of Gondwanaland, and positioned much closer to the Equator. No strata younger than Early Jurassic are found in East Antarctica.

The structure of West Antarctica is that of a deep basin (Byrd Basin) which lies 2500 m below sea level in places. Gneisses, charnockites, schists, amphibolites and diorite batholiths all occur here. These, by and large, are of Pre-Cambrian–Lower Palaeozoic age. The batholiths yield an age of 400–540 million years BP. Younger, basin-filling, sedimentary sequences range in age from Pre-Cambrian to Late Mesozoic, while late Tertiary and Quaternary volcanic rocks exist in Victoria Land, Marie Byrd Land and Graham Land. Mount Erebus is the only volcano which has been active in modern times.

24

The Quaternary Period and ice ages

Ice ages

We are used to regarding the Earth as a world with extremes of climate. Certainly an Eskimo would be very unhappy in central Africa, and a Bushman would be equally unsuited to Greenland. But in fact the full range of temperatures is not nearly so great as might be imagined; a few degrees change in the global temperature can make a surprisingly large difference in

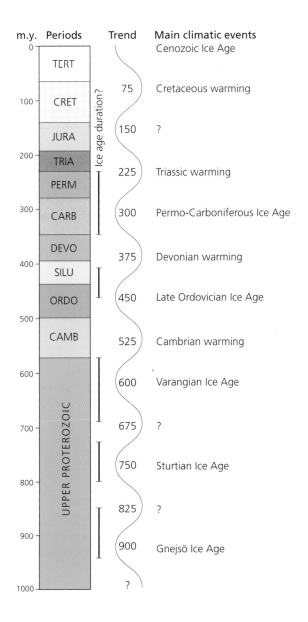

m.y.	Periods		Trend	Main climatic events
0	TERT			Cenozoic Ice Age
100	CRET	Ice age duration?	75	Cretaceous warming
			150	?
200	JURA			
	TRIA		225	Triassic warming
	PERM			
300	CARB		300	Permo-Carboniferous Ice Age
	DEVO		375	Devonian warming
400	SILU			
	ORDO		450	Late Ordovician Ice Age
500	CAMB		525	Cambrian warming
600	UPPER PROTEROZOIC		600	Varangian Ice Age
700			675	?
800			750	Sturtian Ice Age
			825	?
900			900	Gnejsö Ice Age
1000			?	

Major glacial periods. There appears to be a 150-million-year spacing between successive ice ages, although the hypothetical Jurassic ice age apparently did not occur.

Aerial view of crevassed glaciers entering the ocean off east Greenland. Note the lateral moraines and the offshore ice floes. The relative ages of the two glaciers can be deduced from their cross-cutting relationships.

the prevailing climate. If the Arctic and Antarctic ice-sheets melted, the sea would rise by about 70 m; this would affect the distribution of land and sea quite dramatically. Such a change would require a global warming of a mere 5 °C. At present, temperatures are well suited to advanced life over much of the globe, and of the large continents only Antarctica has no indigenous human inhabitants. However, at intervals during its history the Earth has been subject to cold spells, or glaciations. Currently we are living in an interglacial which followed the last of these. The last glacial period began about 25 000 years ago. We call this the Holocene epoch – a name which is really a quite absurd leftover from a misunderstanding of how frequent the glacial–interglacial cycle was during the Quaternary; there is little real reason to afford the last glacial of many a special epoch to itself.

As early as 1795 James Hutton, one of the founders of modern geology, suggested that the Alps had once been covered in much larger masses of ice. The idea of a former cold period became more and more talked about, and was finally demonstrated by the young Swiss geologist Louis Agassiz, in 1830; he presented convincing reasons for believing that the Swiss Alps – and much of northern Europe too – had previously been ice covered. This is not to say that everyone accepted this idea at first, but as soon as the evidence was collected together, the concept of such a large temperature change started to occupy

the minds of geologists and climatologists, as it has done until the present day. Even so, it has taken a surprisingly long time for our understanding of the last ice age to reach its modern level: until the 1970s it was still widely believed that only four (or perhaps five) glacial periods, separated by interglacials, comprised the Pleistocene/Holocene ice age. We now know that there were at least thirty glacials, and that in the distant past there were other major periods of glaciation. At least six have been identified, the oldest having been dated to more than 2300 million years ago.

The Quaternary ice age of the northern hemisphere goes back to around 3 million years BP, well before the start of the Quaternary, while the Antarctic continent has suffered an ice age for at least the past 15 million years. The term glaciation should be qualified: it does not imply that throughout one the climate remained inviolately cold; individual cold periods are always separated by milder interglacials. For instance, during the heyday of the Roman Empire the climate in Europe was appreciably warmer than now (although being technically within the last ice age). There was a colder spell between 400 AD and 800 AD, then another warmer period, which ended about 1200 AD, during which southern Greenland was colonized by the Vikings. This was succeeded by what is often called the Little Ice Age; the climate in northern Europe became severe and the Norse settlements in Greenland were wiped out. Then, after 1780, the temperatures rose again. Since 1940 there has been a slight fall in average temperatures, although it is not likely that this will be anything more than temporary.

It would be quite wrong to suppose that during a glacial period the whole of the Earth is coated in an icy mantle. This has certainly never happened. Although the disposition of continents and oceans was quite different early in Earth history, the evidence shows that the Equatorial regions of the Earth have always been ice-free, even during glacial periods.

Modern Antarctica is covered in ice – and bear in mind that it is larger than the whole of Europe. Greenland, too, is mainly ice-covered. These, however, are the only major continental accumulations of ice at present. The Arctic Ocean is largely ice-covered, though it is only a relatively thin cover of drifting pack-ice. Note that there are extensive land areas in the Arctic not covered by ice, mainly because of the very low precipitation in this region; the Arctic ocean is ever-frozen, does not evaporate and therefore furnishes no precipitation. The same also is true of large areas of continental Siberia, for the same reasons. These polar deserts cover very large areas. In contrast, the damp Pacific winds bring precipitation to Alaska and British Columbia, where there is a considerable snow and ice cover at the present time.

Our present state of knowledge about glaciations, the oceans and climate has come from sophisticated techniques, particularly the measurement of the relative abundances of the isotopes of oxygen by mass spectrometers. Information has also been forthcoming from carbon isotopes, from pollen analysis (palynology) and from deep-sea cores. Together these have provided new insights into a very complex problem.

The Quaternary was also a period of intense volcanic activity, particularly along convergent plate margins. Activity was intense around the Mediterranean and Aegean Seas: the Hellenic arc, Etna and the Aeolian Islands, Santorini, Vesuvius and the Phlegraean Fields being cases in point. Their eruptions pumped enormous quantities of volcanic dust and ash into the atmosphere. The effects on the world's climate are still being studied; they were not inconsiderable. First I must explain the value of oxygen isotopes in geology, particularly to the glaciologist.

Oxygen isotope ratios measured on the calcareous shells of small oceanic organisms. This graph, for the Quaternary period, shows the large number of climatic fluctuations – 60 in all – since the beginning of the northern hemisphere glaciation.

Oxygen isotopes and the last glaciation

There are two stable oxygen isotopes, the lighter one, ^{16}O being by far the more abundant. When a marine organism takes up oxygen in the form of carbon dioxide or carbonate, it builds this into its shell or carapace. In so doing it acquires both ^{16}O and ^{18}O in the proportions they are found in sea water at that time. When the organism dies, the $^{18}O/^{16}O$ ratio is preserved in the remains, where it can be measured with great accuracy.

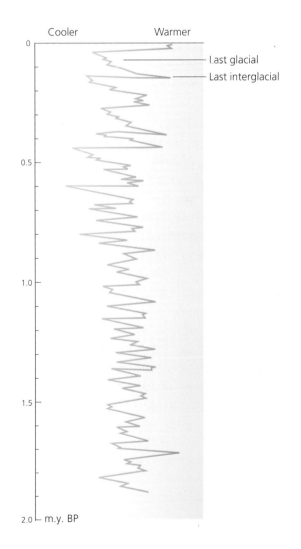

During normal conditions, the lighter isotope evaporates from sea-water more readily than the heavier one, with the result that atmospheric moisture and rain are both enriched in ^{16}O. In consequence a tiny excess of ^{18}O relative to its normal values with respect to ^{16}O would be left in the sea if the evaporated moisture were not returned to the oceans in rain and via rivers, there to be stirred and homogenized by waves and currents. Thus the ratio remains the same everywhere. Things are rather different during a glacial period however, because the evaporated sea-water becomes locked up in ice – together with its excess of ^{16}O – while the ocean has a higher $^{18}O/^{16}O$ ratio than it would have during an interglacial. Calcareous organisms living during an interglacial will, therefore, record the change in oxygen isotope ratio. In consequence if different layers of a deep-sea core are analysed, the oscillations in oxygen isotope ratio pick out the alternating glacial and interglacial periods.

There have been more than 60 major fluctuations since the Quaternary ice age began, more than 2 million years ago. It is also possible to recognize a distinct evolutionary trend: during the early stages the glacials and interglacials were of roughly equal duration, each pair lasting between 40 000 and 50 000 years; the temperature difference between the warmest and coldest was also quite small. Things changed around one million years ago: fewer but longer glacials developed, each being upwards of 100 000 years long. Furthermore, they were separated by shorter interglacials. The oxygen isotope ratios indicated a sharp contrast between the ice cover, the $^{18}O/^{16}O$ ratio rising to nearly twice its original during the glacial episodes.

The oxygen data mainly record the changes in ice cover over the Earth and, since the Antarctic cover has changed little, what we are seeing are the changes which affected the northern hemisphere during the past 2 million years or so. It is, of course,

the last interglacial–glacial cycle about which we know most, largely because no new ice has advanced to eradicate traces of the evidence. It is this last 125 000 years which we will now study.

The character of the last glaciation

Between 25 000 and 70 000 years ago, ice-sheets covered extensive areas of North America and Europe, but were less strongly developed in Siberia. It is believed that a 3-km-thick dome of ice covered much of Canada, extending southwards to New York and the Ohio River valley. In the west, ice covered the northern Rockies and the Cascades. Scandinavia also was covered in ice, and from this collecting ground it spread southwards over Denmark, Germany, Poland and Russia; smaller ice-sheets rested on the Alps, Carpathians and Pyrenees. Scotland and Wales had their own British icecap probably never connected with the larger European one. While the Antarctic ice-sheet was little different from today – although the ice shelf extended much further north – the ice caps of South America, Africa and Australia were far more extensive. Even Hawaii appears to have been ice-covered! At the peak of glaciation it is calculated that 40 km³ of ice lay on the land.

Conditions peripheral to the ice-sheets were not as severe as might at first be imagined. Pollen analysis reveals that forests thrived not far beyond the edge of the ice in North America, implying that summer days were long and bright enough to allow such growth, i.e. conditions were not truly 'polar'. The extent of the ice on the continents was, latitude-for-latitude, greater than that on the oceans. This is due to the way in which ice behaves on a solid surface: it spreads laterally, under its own weight. This is why the ice-sheets extended over such wide areas of the northern hemisphere. In contrast, the essentially sea-covered southern hemisphere saw a less extensive cover of ice. The deep oceans surrounding the Antarctic, in partic-

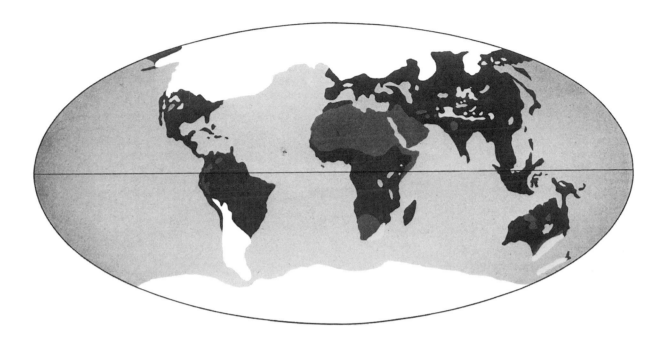

The extent of the ice during the last major glaciation (about 18 000 years BP). Dark grey areas = land covered with forest, grassland and steppe; mid-grey areas = land covered by sandy desert and snow-covered forest; light grey areas = ice-free seas and lakes; white areas = ice and snow.

The cone of Gorelyy volcano in Kamchatka, one of the most active volcanic regions of the world. It is a northern extension of the Pacific's Ring of Fire. This part of eastern Asia, although well south of the Arctic Circle, is under an ice cover for most of the year. It is a reminder of what much of the northern hemisphere looked like during the last ice age. [Photo courtesy of NASA]

ular, meant that the peripheral ice, on reaching deep water, simply broke away, forming floes and icebergs.

This brings me to the topic of the oceans: since the world is two thirds ocean covered, what effects did the oceans have on climate and glaciation? Luckily we have another helpful item here: tiny calcareous and siliceous plankton live at or close to the surface of the sea and record the conditions in the sea in which they lived. Their remains in deep-ocean sediment cores can reveal information about salinity, temperature and even nutrient levels in the glacial oceans. For instance, if we go back to 18 000 years BP, the ocean was frozen as far south as Britain and full of pack ice and floes as far south as Spain and Cape Hatteras. For this reason, the Gulf Stream – which today warms the shores of the Hebrides – flowed straight across the Atlantic from Florida to the Azores and warmed Africa alone. The sea temperature in Spain would have been only around 7 °C – not terribly inviting to the package holidaymaker!

Equatorial sea temperatures were considerably lower than today's, but in the mid-latitudes there was little change in temperature. California and the Indonesian Archipelago would have enjoyed very sim-

ilar temperatures to those of today; however, rainfall levels would have been considerably higher, because high pressure over the northern ice-cap deflected Pacific winter storms much further south than at present. Most of the world was, however, drier, largely because the coolness of the ocean water led to reduced evaporation and thus lower average precipitation.

On land, temperatures were cooler everywhere. Palynological analysis – the study of fossil pollen – quite clearly shows that on the northern continents, beyond the tundra zone, a forest of birch, pine and spruce thrived, much as it does today in Canada. The temperate broad-leafed forests, at this time, would have extended further south, for instance to Florida and the Gulf Coast, and to southern Europe and North Africa. In the southern hemisphere, lower temperatures and less rainfall meant that the area covered by tropical rainforest diminished. Much of what now is lush forest, then was brown savannah.

Changing levels of land and sea

Each time the ice extended, so the level of the ocean

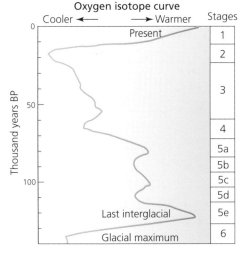

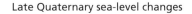

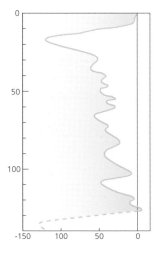

Oxygen isotopes and sea-level changes for the past 140 000 years. This covers the period from late in the previous glaciation until the present time.

dropped; each time the ice melted, so the sea level rose, extending over the coastal plains of the continental peripheries. Such worldwide changes in sea level are said to be *eustatic* – they occur at the same time everywhere. During the peak of the last glaciation (18 000 years BP) the sea level was between 100 and 150 m lower than now. If all of today's Antarctic and Arctic ice were to melt, the level would rise by about 70 m.

During the early days of research, many studies were made of ancient shorelines, left by retreating ice-caps during interglacial periods. Radioacarbon dating of wood or shell remains ought to lead to an understanding of the rise-and-fall history. Unfortunately it did not, things turned out to be far more complex than predicted.

This was largely due to the fact that when the crustal rocks are either loaded or unloaded with their ice burden, they respond isostatically, there being a distinct time lag between the removal of the ice burden and the subsequent isostatic rebound. What had not been realized was that the oceans too, experienced such changes. In this case the weighing-down was due to the water in the oceans: during interglacials there would be more water in the ocean basins, so their floors became relatively depressed. During glacial periods, this weight was lessened, and isostatic changes occurred. As a rule of thumb, every 100-m-worth of water removed from the ocean will cause the sea-floor to rise by 30 m, while each 100-m-thick layer of ice will depress the continental crust by around 27 m. The see-sawing of levels turns out to be very complicated indeed, and while calculating the volume of ice once present is not too difficult, numerous corrections have to be applied before it is possible to arrive at a reasonable figure for either the rise or fall in sea level as glaciations proceed.

Going back to the last glaciation – it appears that the sea level was about 120 m lower than at present. For this reason the continental shelves, which are on average less than 150 m below sea-level, would have emerged from the waters and extended the continental area considerably; the geography of the world would have looked quite different. Only where the shelves were narrow, for instance along the western Americas, would little change have occurred. One of two suprising things would have occurred: the Indonesian Archipelago – currently separated from mainland Asia by the broad but shallow Java Sea – would have been joined to the mainland; the North Sea would have been land, as would a 200-km-wide plain bordering the Gulf of Mexico.

About 17 000 years ago, the ice-sheet finally began to melt; sea level rose slowly at first, then accelerated until, around 12 000 years BP, in a period of 1000 years it rose by 24 m! Following this there was a quick return to colder conditions during what has been called the Younger Dryas (about 10 500 years BP), and the sea-level rise was halted for a while; thereafter it speeded up once more and the present level was attained. It is estimated that a further 7–10 m rise is in store, possibly from the western part of the Antarctic ice-sheet.

Onset and decline of an ice age

There is much that has been written which suggests that the last glacial maximum was typical of the Pleistocene as a whole. This is not so, for conditions were much of the time far less severe. Archaeologists' propensity to have early hominids in a Europe gripped by ice is far from the truth. It is generally believed that the last interglacial occurred around 125 000 years BP; isotopes of oxygen tell us this. Furthermore, isotopic analysis of cores which sampled the entire thickness of the Eastern Antarctic and Greenland ice-sheets show that ice which formed during this last interglacial still exists at the bottom of the ice-sheets at those two localities. It is likely, therefore, that the high sea level was in large part due to

The limits of ice-sheets during the peak of the last glaciation: (left) North America; (right) Scandinavia. The USA, in particular, was much more ice-free than most people believe.

the melting of Western Antarctic ice.

Following on from this, around 110 000 years BP, there were two phases of cooler conditions interspersed by a slight amelioration. The first major glaciation arrived 75 000 years ago – pollen analysis tells us that forests disappeared from mid-latitudes, being replaced by arctic-type tundra. The climate then ameliorated once more before the true glacial maximum arrived, around 25 000 years ago; it reached its peak 18 000 years BP.

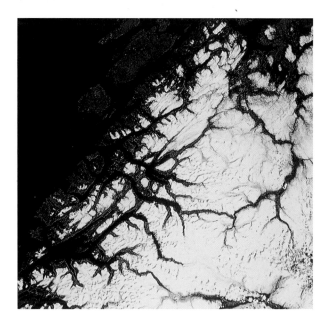

At the height of the last major interglacial, sea level was around 6 m higher than now; at the peak of the last glaciation it plummeted 120 m below the modern level. Until quite recently it was not appreciated how, shortly after the glacial peak, the ice-sheets of the world began to melt. For instance, a mere 1000 years afterwards, the North American ice-sheet retreated at the rate of 1000 m per year; the eastern ice-sheet withdrew from Ohio down to Hudson Bay – a distance of 2000 km – in under 6000 years and had gone entirely 7000 years ago. The decline of the glacial period was strikingly rapid, the ice retreating from Britain, Denmark and northern Germany by 11 000 years BP. Indeed, ice core data show that climatic change is far more rapid than we once thought, sharp changes from warm to bitterly cold occurring in just a few hundred years.

During what is known as the Atlantic period (8200–5900 years BP) the ice had disappeared almost completely from the northern continents; Greenland was the only exception. Deciduous forests then dominated the continents, the average temperature being

Part of the glacier-carved coast and ice-cap of Norway, south of the city of Tromsø (top margin of photo). All of Scandinavia was covered by ice at glacial maximum, but for much of the glacial cycle the ice occupied only the central highlands of Norway. The Kjølen Mountains – seen in this image – rise to 2400 m and are a part of the great Caledonian Mountain chain. [Landsat image]

Aerial view of Swiss Alps, showing straight-sided, U-shaped valley scoured by Pleistocene ice-sheets. Straightening of valley sides is a feature of glaciation, together with the cutting back along joints and faults to produce sharp-sided aretes and triangular-shaped peaks. Currently ice is restricted to the higher peaks and cirques.

U-shaped valley indicative of past glaciation of a region in temperate latitudes. This lush valley, Dyffryn Penllyn, is located near the town of Bala in central Wales and was just one of many which during the peak of glaciation would have been filled by a glacier. The U-shape and lack of lateral shoulders is typical of glacial modification of a pre-existing valley.

Boulder clay deposits near Aberystwyth, Wales.

perhaps 1 °C higher than today. The evidence shows that rainfall over the Sahara and East Africa was considerably higher than at present, and African lake levels, which had fallen by several hundred metres during the last glacial, were significantly higher than now. The modern climate appears to have been established around 2250 years ago.

Moving ahead in time, between 950 and 1250 AD, Greenland became a green and pleasant land, as a warm period set in. Scandinavian settlers travelled there at that time. However, this was followed by what is known as the Little Ice Age, which reached its peak in the seventeenth century and waned during the eighteenth. The Thames frequently froze solid at this time, as old engravings show. Further slight oscillations have occurred, but the most marked of these is that which started in the 1980s: it is getting warmer all the time. Remembering how climatic changes happened so quickly in the recent past, should spur us to examine very closely the greenhouse conditions which humankind's ineptitude in managing itself may be bringing on.

What causes an ice age?

The fact that the Earth has experienced glaciations throughout its long history has been known for more than a century and a half, but we are still uncertain as to the reasons. Many theories have been proposed, some of them decidedly far-fetched. One seemingly profitable line of enquiry is an astronomical one: have there been changes in the output of the Sun? On the whole the Sun is a perfectly normal, well-behaved star; admittedly it is not always equally active, there being sunspot and Solar Wind maxima every 11 years or so. However, there seems to be no real connection between sunspot activity and the climate and we cannot even be sure that the cycle is permanent. Additionally, we have no evidence that our star has gone through long-term changes since it stabilized and joined the Main Sequence.

An entirely different proposal was made by the astronomer John Herschel in 1830, not long after the existence of ice ages had been demonstrated. In 1867 a similar idea was pursued by James Croll. The theory was refined and republished by the Yugoslav scientist

M. Milankovich, and is usually known as the Milankovich Theory. Over long periods there are changes in both the Earth's orbit and its axial inclination. We call this *precession*. Starting first with the orbit: the Earth's path around the Sun is an ellipse with a small eccentricity that varies by about 6 per cent over a period of around 100 000 years. At the present day, the Sun is at perihelion on 3 January and at aphelion on 4 July; this means that the winters are not as cold as they might be, while the summers are cooler. The reverse is true at the other end of the cycle, giving a slight change to the climatic cycle.

The rotational axis of the Earth is tilted to the plane of the orbit, currently at an angle of 23.4°; because of this the northern hemisphere receives more solar radiation during the part of the year when it faces the Sun, while the southern hemisphere enjoys more warmth alternately. Now the obliquity changes between 21.8° and 24.4°, over a period of 41 000 years. Consequently the length of the day at high latitudes changes as the cycle progresses, and so therefore does the amount of radiation received there.

The Earth's axis also wobbles over a period of 22 000 years, describing a cone against the star background as the cycle proceeds. This effect shifts the dates of the solstices and equinoxes clockwise around the orbit. Currently, the northern winter begins just before the Earth is at perihelion; however, 11 000 years ago it began on 21 June!

These precessional cycles must have some cumulative effects. As far as the total amount of solar radiation received by the Earth is concerned, only the eccentricity has real influence. This amounts to a change in heat of only 0.3 per cent – surely too little to be any kind of explanation. However, what we have to take on board is the changing heat distribution with latitude and season that obliquity and precession bring about. For instance, at high latitudes, temperature changes in winter have little impact since it is always cool enough for snow and ice to accumulate. In contrast, if the summers are cooler than now, less snow melts, with the result that at the onset of the ensuing winter, the area of snow has not diminished by as much as before. Consequently the area of the snow slowly increases with time. So, the Earth is in a position favourable for ice formation when the Sun is distant in summer and the tilt is small.

Although such variations seem quite small, they can be tested by looking carefully at the oxygen isotope data, pollen data and the rock record generally. Taking the oxygen data: in 1979 a group of scientists led by John Imbrie of Brown University, Rhode Island, showed quite conclusively that the oxygen isotope ratios vary over time as the sum of three curves which have periods of 100 000, 41 000 and 22 000 years. Surely this cannot be mere coincidence, it must give credence to the Milankovich idea?

Currently a large number of scientists accept this precessional theory, even though it has its shortcomings and not all of the details have been finally worked out. One of the obvious problems is accounting for the seemingly rather abrupt changes from glacial to interglacial within each cycle. The precessional changes are surely rather slow to account for this? However, it may be that other components of the climatic cycle, for instance, moisture supply, greenhouse gases, global albedo and amount of volcanic eruptives, may all play a part. Indeed there is some evidence to suggest that the annual monsoonal winds are also affected by the orbital geometry. Then again, amounts of carbon dioxide measured from sediments in deep-sea cores were, during interglacial periods, only a little below current values while they dropped significantly during glacial episodes. Perhaps the greenhouse effect plays some part too, but is it cause or effect? The answer to this will have to await futher research.

25

Early man

Introduction

Despite all the arguments which have raged over the years, we are still uncertian about the origin of mankind. There are also several misconceptions. Many people genuinely believe that the great naturalist Charles Darwin suggested that men are descended from monkeys. Of course, this is sheer nonsense; Darwin would have been the last to suggest anything of the sort. What he did say, with perfect accuracy, is that men, monkeys and apes have common ancestry.

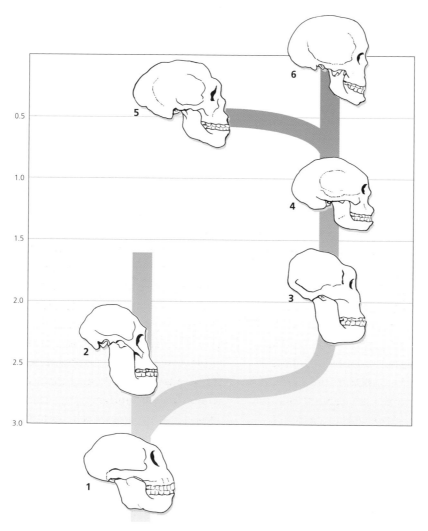

Man's ancestry can best be traced through skulls which occur as remains in different parts of the world. They enable us to measure the size of the brain. (1) *Ramepithecus*; (2) *Australopithecus afarensis*; (3) *Homo abilis*; (4) *Homo erectus*; (5) *Homo sapiens neanderthalis*; (6) *Homo sapiens sapiens*.

If we could trace our descent sufficiently far back, we would come to the small, largely tree-living primates of more than 65 million years ago.

Moreover, even today there are opponents to the idea of evolution. In Britain, the Evolutionist Protest Movement is regarded with amused tolerance, but in Arkansas and other parts of the USA there is a move to restore the teaching of creationism: the idea that man appeared quite suddenly, in his modern form, by divine agency. The reasons for this belief are religious (or, rather, pseudo-religious) but the movement does show that still today scientists have to battle with prejudice as well as ignorance.

We must also try to decide exactly what we mean by 'man'. The search for the so-called Missing Link, a creature that is half-human and half-ape, still goes on, but fossils pre-dating the last Ice Age are rather rare and fragmentary, and the evidence is inconclusive. (There was, of course, the famous Piltdown Man, which was dug up in a gravel pit in Sussex and was classed as a true Missing Link inasmuch as it had a jaw like that of an ape, but a skull that certainly was human. Alas, *Eoanthropus*, or Dawn Man, proved to be a fake. The skull was of moderate age, while the jaw was that of a modern orang-utan, carefully treated so as to confuse the experts.)

Certainly there was fairly rapid evolution around 50 million years ago. During the Oligocene epoch, which extended from 38 million to 26 million years ago, there were many species of monkeys and apes, classified as hominoids. There is no chance that man is anything like as old as this, but it seems that hominoids, from which we are descended, branched off from all other primates around 35 million years BP. Then perhaps as long as 10 million years ago, there was another development: a new branch of hominoids, called the dryopithecines, became distinctly separate from the rest. These were the ancestors of modern gorillas and chimpanzees. The hominoids of this new branch could walk erect, had a small mon-key-like brain and may even have used primitive tools, although here opinions differ. However, in terms of their jaw formation and dentition they were ape-like, or even man-like.

It is most unlikely that they could talk in the true sense of the word (although even here we must be cautious in view of recent experiments which indicate that dolphins have something which is remarkably akin to a rudimentary language). They were not hunters in an organized sense, although they must have killed animals for food. The glimmerings of civilization, such as the making of fire, were completely beyond them.

The Neogene developments

The Neogene (which commenced about 26 million years BP, and includes the Miocene and Pliocene epochs) saw the major evolutionary steps that eventually led to the appearance of modern humans, or *Homo sapiens*. The lineage which produced modern humans can be traced back to African primates, the remains of which have been dug up in east Africa. The earliest was *Aegyptopithecus*, a small tree-dwelling primate that lived in Egypt 28 million years ago, during the Oligocene epoch. Next came *Dryopithecus*; this lived in Africa between 20 and 15 million years BP (early–middle Miocene) and expanded into Eurasia about 14 million years ago, when the Arabian craton collided with Asia. The foot bones of this creature indicate that it did not hang from trees but walked on land, possibly upright.

Kenyapithecus dates back to 12 million years. This was another advanced primate from Africa whose remains have been found alongside piles of bones that had been crushed with stones that had been imported into this region of East Africa. A branch of this family then migrated to Asia; this was called *Ramapithecus*; it had human-like teeth. Regrettably there is then a 6-million year gap in the fossil record

The head of *Australopithecus*.

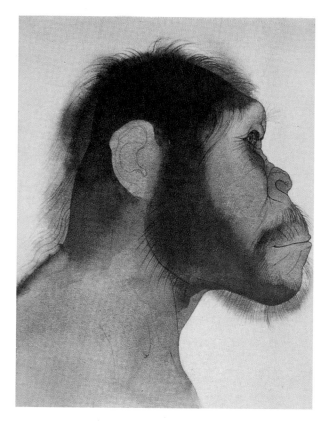

before the earliest Hominidae appear. This family includes two fossil genera – *Australopithecus* and *Paranthropus*, and the living genus *Homo*. This brings us to *Australopithecus* and the startling discoveries made in the Olduvai Gorge.

From Olduvai to *Homo sapiens*

It has been said that the Olduvai Gorge, in Tanzania, is one of the most interesting places in the world. It was here, in 1924, that Raymond Dart found the first evidence for the hominid (truly manlike), *Australopithecus*, dating back to the Pliocene epoch. *Australopithecus* was of a higher order than any ape. With a massive head, projecting jaw and small cranial cap, it must have been a forbidding figure by our standards, but it was advanced enough to use stone implements.

Other finds made at Olduvai include remains of *Homo erectus*. This Middle Pleistocene hominid was widespread in Europe, Africa and Asia and marks yet another major step forward in evolution. The remains range in age between 1.6 million and 400 000 years BP. This hominid had a brain that was over three-quarters that of a modern human. It is not entirely certain whether *Homo erectus* was descended directly from *Australopithecus*; for instance, a skull has been discovered at Lake Rudolph, in Kenya, which is estimated to be three million years old and not similar to *Australopithecus* at all; so there may well have been more than one line of hominid evolution.

The earliest species of Australopithecus, *Australopithecus afarensis*, was found in Ethiopia and lived between 4 and 3 million years ago. It stood about 1 m high, weighed about 30 kg, and its brain size was roughly half that of a modern human. *Australopithecus africanus* was somewhat more advanced and lived between 3 and 2 million years BP. Its territory was eastern and southern Africa, and it had a brain size roughly seven tenths that of a modern human being.

The genus *Paranthropus* evolved in parallel with *Australopithecus*. The former was a heavier-built creature but had a comparatively small brain size; it was a vegetarian. Its remains also were found in Africa, and they range in age between 2.5 and 1.5 million years BP. For some reason it had no descendants.

Homo erectus flourished in China and Europe as well as Africa, and apparently was well established during the earlier part of the Pleistocene epoch. Some specimens have been found associated with *Australopithecus* in South Africa, and so it seems likely that the smaller-brained genus was only slowly replaced by its larger-brained descendant, the two co-existing for a long time.

Another find of special importance came from a cave near Zhoukoudian, 48 km southwest of Peking. This cave was inhabited for a very long time. Peking Man (the name established before the transliteration Beijing for the name of the Chinese city was officially adopted) first lived there about 460 000 years ago, and did not vacate it until 23 000 years BP, when the cave became filled with rubble and sediment. Peking Man appears to belong to one of the same species (*Homo*

Fresco from the Minoan 'West House' at Thira, showing a fisherman. Dated at about 1500 BC. This site was destroyed by the huge eruption of Thira, which occurred shortly after this time. This confirms the suspicion that humankind's presence on this planet is, at best, subject to the whim of natural forces and, at worst, a temporary affair. [Athens, National Museum]

erectus) as came from Olduvai. He was a hunter-gatherer and certainly he could make use of fire. Among the implements found in the cave are choppers and scrapers made from materials such as rock crystal, flint and sandstone.

The earliest specimens of *Homo sapiens* are Swanscombe Man (from England) and Stuttgart Man (from Germany), both of which go back to between 250 000 and 200 000 years BP. The brow ridges of both specimens are like those of *Homo erectus*, but the back of the skull is rounded, and more modern in type than that of Neanderthal Man – who appeared later, about 120 000 years ago, at or just before the start of the last glaciation

Neanderthal Man used to pose a problem. He was sturdy in build, with massive brow ridges; he was a hunter, and he used bone and wood to make tools. He may have clothed himself with animal skins, and even buried his dead. He is the best documented example of a heavily-built race which lived about 120 000 years ago. About 35 000 years BP, when the last glaciation was at its severest, Neanderthal Man disappeared. Whether the race, which lived in both Europe and Asia, was wiped out by the arrival of the far more advanced Cro-Magnon Man, we do not know. What we now believe, however, is that both the Neanderthal and Cro-Magnon Man are too similar to modern man to merit a separate species, and that well before the end of the Pleistocene epoch (about 500 000 years BP), *Homo sapiens* had become dominant

From archaeology to history

The last glaciation ended about 10 000 years ago: or, in historical dating 8000 BC. It is probable that true language was developed at an early stage in the existence of *Homo sapiens*. He was a hunter on an organized scale, and as the ice began to retreat he followed the herds northwards.

Mammoths were one source of food – no doubt a

vitally important one – as well as providing skins for clothes and even primitive housing. The fact that the mammoth became extinct can have been due to nothing but its ruthless slaughter by these early hunters. Well before the end of the last glacial peak, too, we have excellent evidence that peoples who lived on the edges of glaciers in what is now Czechoslovakia and Russia, built huts from poles covered with animal skins sewn together. Men knew not only how to control fire, but also how to make it. The huts were movable, but it may not be going too far to speak of 'villages', even though they were not permanent.

With the development of Neolithic cultures, which began about 9000 years ago, man became a farmer; with the development of agriculture came the first use of domesticated animals. The rate of progress was naturally different in different parts of the world, but by 5000 years ago, humanity had spread from its places of origin across to Australia and the Americas.

Up to now we have been dealing with archaeology. Historians do not really come into the picture until we can identify the first true civilizations; the borderline is by no means clear-cut, but it may be somewhere between 6000 and 4000 BC. For instance, the Sumerians flourished by 4000 BC, and Eridu, 19 km from Ur of the Chaldees, seems to have been a permanent settlement with reed huts and mud-brick houses. Cuneiform writing had been developed before 3000 BC, and by then Egypt had been unified; according to tradition the first king of Egypt was Menes, who had a long reign that ended when he was killed by a hippopotamus. The Old Kingdom, between about 2700 and 2150 BC, was the age of pyramid building, of the amazing decorated tombs of the Valley of the Kings, and of Luxor; nobody who has seen these immense structures can doubt that Egyptian civilization had reached a very high level indeed.

Other parts of the world suppported civilizations; in India from at least 2000 BC, in Greece somewhat earlier. By 2500 BC the Cretan civilization was well advanced and lasted for over a thousand years. Apparently its demise was hastened by the enormous explosive eruption of Thira, around 1500 BC. It was almost certainly the disaster which gave rise to the legend of Atlantis. Henceforth the story belongs not to geologists, but to historians.

26

The Earth now

Natural disasters and hazards

Earthquakes, volcanic eruptions and hurricanes are perhaps the three phenomena which spring to mind when natural hazards and disasters are mentioned. Recent volcanic disasters in the USA, Mexico, the Phillipines, Indonesia and the Caribbean have certainly focussed both public and scientific attention on the problems of forecasting and alleviating volcanic hazards. The recent major earthquakes in California, Japan and Columbia have maintained the high profile of this kind of hazard too. Hurricanes recently have caused much damage in the Caribbean and the western USA, while flooding has wreaked terrible havoc in Bangladesh. While these are unquestionably the major hazards, marine erosion, deforestation and bad land usage all contribute to disasters, even if over quite small areas. The recent increased awareness of natural hazards prompted the US National Academy of Sciences to propose the 1990s as the International Decade for Natural Hazard Reduction.

There has been much recent work on the study of the environmental impacts of major eruptions on the Earth's atmosphere and world climate. Indeed, as we have seen, it has even been suggested that major eruptions can contribute to glaciations. New geophysical and geochemical techniques have been developed which allow scientists to monitor volcanic activity better, to gain a greater understanding of how volcanoes actually work, and these are slowly leading towards the situation where predictions about volcano unrest can become more accurate. Similar developments have been taking place in the field of seismicity, and considerable research has been ongoing in the field of structural engineering, with a view to minimizing structural damage (and loss of life) in earthquake-prone areas of the world.

Of course, certain areas of the world are more prone to some types of hazard than others. Indonesia and the islands of the southwest Pacific are located

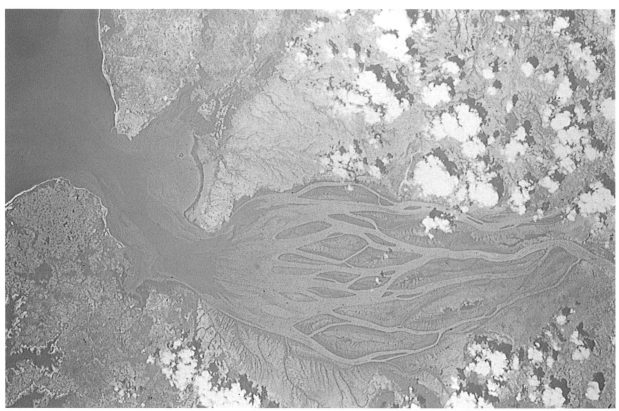

Space Shuttle image of the sediment-laden drainages of the Betsiboka River, Madagascar. The red lateritic soils which form the sediment seen in this photo represent a disaster for the island. The river's load is, regrettably, due to uncontrolled felling and clearing of Madagascar's forests which has resulted in rampant soil erosion.

where a complex of lithospheric plates meet. Subduction volcanoes and earthquakes make this a particularly violent sector of the Earth's surface. Add to this the tropical storms which periodically sweep through the region, bringing down mudslides and triggering huge landslips, and you have an area of concentrated natural hazards. Nearly the whole of Bangladesh occupies the floodplain and delta of the Brahmaputra and Ganges rivers, and is subject to serious flooding.

The atmosphere is tipped into action by extremes of temperature. This may trigger off winds which develop into tropical storms or hurricanes of immense power. A large hurricane, for instance, can generate as much energy in 24 hours as the USA uses in electricity in one entire year! Cyclones annually affect the western coasts of the Pacific, Southeast Asia, the Indian subcontinent, the Caribbean and southeastern USA and the southeastern coast of Africa.

Man also interferes with nature and can directly or indirectly contribute to nature's propensity for hazard. Deforestation often leads to soil erosion and gullying, to increased instability on steep slopes and a change in the pattern of drainage and flooding. It also contributes to a change in the amounts of carbon dioxide involved in the Earth's atmospheric cycle. The burning of fossil fuels and use of certain chemicals in aerosol sprays have both been cited as contributors to the production of holes in the planet's protective ozone layer. Hazards are all around us.

A 25-year sample of natural disasters revealed that 39 per cent of deaths were associated with floods, 36 per cent with hurricanes and typhoons; next came earthquakes – (13 per cent), gales and thunderstorms (5 per cent), then volcanic eruptions (2 per cent). Tornadoes, snowstorms, heatwaves, cold snaps and landslides each accounted for about 1 per cent of deaths.

In this chapter are presented details of some of the relatively recent geological disasters which have prompted scientists to increase their knowledge as quickly as possible, enabling them to give the best advice possible to those seeking to minimize destruction and loss of life on the ground.

Krakatau and all that

The volcano Krakatau – often called Krakatoa – lies in the Sunda Strait between the Indonesian islands of Java and Sumatra. In about 416 AD an eruption caused it to collapse, forming a 8-km-wide caldera. The islands of Krakatau, Verlaten and Lang are the remnants of this ancient volcano. There was also a fairly major eruption in 1680, when considerable quantities of dacitic debris were expelled. The eruption and subsequent collapse of this modified Krakatau structure in 1883 produced one of the largest explosions on Earth in recorded time. It destroyed much of Krakatau Island, leaving only a small remnant; and was one of the Earth's most infamous natural catastrophes.

Following on from three months of quite small, intermittent eruptions and a single day of larger

Recent eruption (1997) from the vent of Anak Krakatau [Photo: Mike Lyvers]

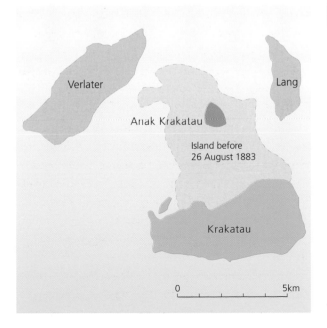

The general physiography of Krakatau, Sunda Strait, Indonesia.

blasts, 36 417 local people worked, slept, ate and drank, totally unaware that they shortly would become victims of one of Nature's greatest tragedies. Then, on 27 August 1883 a cataclysmic explosion rocked the world. It raised an ash cloud 50 km into the atmosphere, and a sound wave which was heard as far away as Rodriguez Island, 4653 km distant across the Indian Ocean and also in Australia. Barographs around the world registered the immense shock wave too, some of them as many as seven times.

The ash from the eruption plume fell on Singapore, 840 km to the north, on the Cocos Islands, 1155 km to the southwest, and on ships as far away as 6076 km to the west-northwest. Darkness fell over the Sunda Strait from 11 am on the 27 August until dawn the following morning. Giant waves reached heights 40 m above sea level, devastating everything in their path and hurling coral blocks weighing up to 600 tons ashore. A Dutch ship was hurled inland as if a piece of jetsam.

Three years prior to this event, R. D. M. Verbeek, a Dutch geologist, had made a study of the pumice from the 1680 eruption, noting that its high silica content was such as to have potential for very explosive eruptions. On the day prior to the great eruption, Verbeek was at home in the Javanese highlands. On Sunday, 26 August he heard a low and distant rumbling; he had no doubt from whence this came and made a note in his diary of the time – 1.06 pm.

At Anyer, a town 45 km across the Sunda Strait from the volcano, telegraph master Schruit, was sitting down to his Sunday dinner, feeling pleased that he had at last found a suitable house in which to bring his family, from whom he had been separated since his appointment to the post. The new house was close to the beach, with an extensive view over the sea. After his meal he strolled on to the verandah and noted a Government ship, the *Loudon*, steaming up the Strait in a great hurry. Looking to his left he suddenly saw a column of steam rising above the island;

it billowed up like thousands of white balloons and he instinctively knew that this was something very serious indeed. The surface of the sea was alternately rising and falling, and the eruption cloud rolled across the Straits encompassing them in almost total darkness. He quickly groped his way to the telegraph office and ordered his assistant to send a warning to (what is now) Jakarta. It was 2 pm. He got his message through but, just a few minutes after 2 o'clock the line went dead. Anyer had telegraphed its last message. A day later thousands were dead and all the villages along Java's western coast were piles of rubble and corpses. Mrs Schruit never saw her house again. In all 165 coastal villages were destroyed.

It was the collapse of the volcano which caused most destruction, not the blast. The wall of water rolled across the ocean, reaching mountainous proportions as it reached the Javanese shore. It was preceded and followed by a tremendous wind. Rising higher than the tallest palm tree, it destroyed anything in its path. Where it was funnelled into a narrow bore at the Sunda Strait's eastern entrance, it was 150 m high; it escaped more freely to the west, sweeping across the Indian Ocean, touching Cape Horn, and two days later, raised sea level in the English Channel by 5 cm. Tide gauges recorded the tsunami's passage across the Earth: it reached Aden in 12 hours (a distance of 3800 nautical miles).

Floating pumice was so thick that it completely clogged the propellers of ships and made many harbours unusable. Rafts of floating pumice were locally thick enough to support the weight of men and trees. Almost certainly other biological passengers used these to travel across the Indian Ocean, taking about 10 months to reach their destination.

When the eruption ended, only one-third of Krakatau remained above sea level, and new islands of steaming pumice and ash lay to the north where the sea previously had been 36 m deep. Blue and green Suns were observed as fine ash and volcanic dust formed aerosols 50 km into the stratosphere. These encircled the Equator in 13 days. Three months after the eruption these materials had spread to higher latitudes causing vivid red sunsets; they were so vivid that fire engines were called out in New York! These unusual sunsets continued for three years. The volcanic dust which created these spectacular atmospheric effects also acted as a solar radiation filter. Global temperatures were lowered by as much as 1.2 °C in the year following the catastrophe.

Since 1927, small eruptions have been frequent and have slowly constructed a new island, Anak Krakatau (Child of Krakatau). In January 1960 a team of scientists visited Anak Krakatau and recorded renewed explosive activity involving pyroclasts ranging in size from dust to large boulders. At this time the island had a minimum diameter of 1.5 km and was 166 m high. The crater was 600 m across and contained a growing cinder cone. Explosive, Vulcanian-style, eruptions took place at 0.5–10 minute intervals, the largest hurling ash and lapilli 1200 m into the air.

That cycle of activity commenced in 1959 and ended in 1963. Since that time Anak Krakatau has experienced at least nine active phases, most lasting less than a year. The most recent episode commenced in March 1994 and has continued, on-and-off until the present day.

Mount St Helens

In a report dated 1975, Mount St Helens was cited as the volcano most likely to erupt in the Cascades Province of the western United States. On 18 May 1980, after nearly two months of intermittent earthquakes and steam activity, a violent 9-hour-long eruption sent volcanic ash into the atmosphere, devastated more than 100 km² of countryside, killed thousands of wild animals and left about 50 people dead. Effectively it turned a part of Washington State into a living hell.

Mount St Helens is one of a number of high volcanic cones situated along the Cascades Range which extends from northern California to southern British Columbia. The chain of active or recently-active volcanoes is a part of the belt of volcanicity that encircles the Pacific, and has been nicknamed the 'Ring of Fire'. Even so, activity on the scale witnessed in 1980 had not previously been recorded in the Cascades during the twentieth century; the only other eruption of note having been that of Lassen Peak, in 1914.

Following a series of relatively minor earthquakes, the volcano first emitted steam and ash on 27 March 1980. For the following 7 weeks, periods of quiescence alternated with activity. The number of earthquakes then increased and harmonic tremor set in – a sure sign that magma was rising into the substructure of the mountain. On 23 April, an ominous bulge appeared on the volcano's northeast flank, close to the summit. For the next 21 days this rose by between 1.5 and 2 m per day; it measured 1 km in length by 2 km across, and by the time it had attained its maximum size was 135 m high.

The principal destructive blast originated high on the mountain's north flank, and came with little warning on the morning of 18 May 1980. There were two strong earthquakes centred about 2 km below the summit; the bulge began to shake and an avalanche of mixed rock, ice and mud swept down the volcano, to be followed by a violent sideways blast that sent hot gas and rock fragments into the air at hurricane velocities. The devastation associated with this opening event extended over 10 km from the focus of the eruption. As if this were not enough, following the avalanche, extremely hot torrents of fragmented volcanic material (pyroclastic flows) and mudflows (lahars) poured down the mountainside, entering streams and leaving massive deposits of sediment which completely choked the major river channels of southwest Washington State, including the Columbia River shipping channel, over 50 km distant.

(a)

(b)

(c)

The 18 May 1980 eruption of Mount St Helens: (a) 8.32 am (b) 8.33 am (c) 8.34.20 seconds.

The cloud of volcanic ash sent out by the blast rose over 12 km into the atmosphere and was carried by the prevailing winds in an eastward direction, blanketing in its course vast areas of America as far east as western Montana. During the main eruptive phase, nearly 4 000 000 000 m³ of old mountain and new magmatic material was strewn over the surrounding countryside, including a large volume of ice.

During the first few hours of the eruption, Mount St Helens blew out approximately the same amount of volcanic material as did Mount Vesuvius in the eruption of 79 AD. The press photographer, Reid Blackburn, who had donated his time to monitoring activity on the mountain, was at a camp 10 km away from the explosion. He perished when the blast occurred. The 31-year-old geologist, David Johnston, who only a few days before had made a daring ascent into the summit crater to collect samples of the volcanic gases, also died in the explosion, at an observation post just 5.7 km from the summit. His last words

were: 'Vancouver, this is it! . . . She's going!'. After the event a new crater, 3 km long and 2 km wide, appeared in the mountain's northern flank.

The disaster was not only a human one. The boiling mud which was flushed down into the rivers and streams, made the water so hot that salmon leaped out onto the river banks, dying there in their millions. The avalanches also destroyed thousands of trees which were transported into the rivers and clogged them completely in places. The loss of wildlife also was immense.

Today the mountain is quiet again. The ash-strewn landscape has already been recolonized with vegetation and the animals have also moved back in. It has become America's newest National Park and has a series of excellent trails and a magnificent new visitor centre. Although the devastating eruption which took place does not imply that other Cascades centres will burst into activity, it does serve as a salutary reminder than one day they might. This is why Mount

Simplified fault map of California showing, in heavy black lines, those
faults which have been responsible for major earthquakes since 1836.
Earthquake intensity is shown in parentheses, on the Richter Scale.

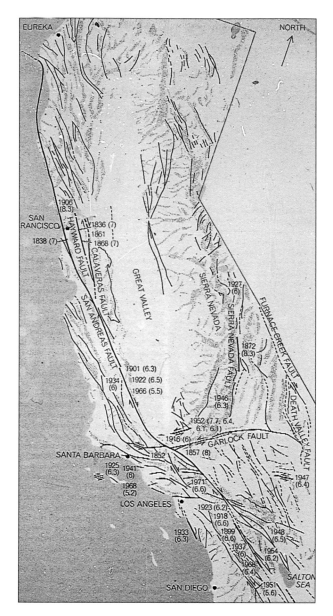

St Helens has become one of the most closely studied
events in recent times. Only by studying such phenomena can we hope to learn enough to make prediction and take precautions against such mighty
hazards.

The San Andreas Fault and Californian earthquakes

The movement of the North American and Pacific
Plates is an ongoing phenomenon. As a result the
Pacific and North American Plates are locked
together for lengthy periods, pressure building up
deep within the Earth's interior. Eventually the crust
gives way, and there is a lateral movement along the
principal line of weakness, the San Andreas Fault.
The most famous documented movement took place
in 1904, but many others have also made the press
headlines, the latest in 1994.

Going back to what has been said earlier, remember that the Pacific is the descendant of the ancient
ocean of Panthalassa but that subsequently all of the
Palaeozoic and most of the Mesozoic oceanic crust
has been subducted, greatly increasing the dimensions of the North American continent. During the
Early Mesozoic, an island arc was driven against the
continent's margin, leaving a suture that can be
observed in the foothills of the Sierra Nevada.
Subduction continued, more material accreting onto
the continental margin, until eventually a line of
andesitic volcanoes rose above deep plutonic intrusions. Then, during the early part of the Cenozoic,
there was a change in the plate tectonic pattern. The
subduction zone was replaced by a transform fault
(the San Andreas Fault), which connects the subduction zone situated off Central America with a spreading axis located off the coast of Oregon. Along the
line of this fault, a slice of California bearing the cities of San Francisco, Los Angeles and San Diego is
drifting slowly towards the Gulf of Alaska at a rate of

around 45 mm per year. About two thirds of this
movement is taken up along the San Andreas Fault
system, which includes the San Andreas Fault itself,
together with some parallel faults, i.e. the San
Jacinto, Elsinore and Imperial Faults.

However, this is not the whole story. In southern
California the Pacific Plate is moving northwestwards
past the North American Plate but whereas in central
and northern California the movement is parallel to
the San Andreas line, here, between the southern end
of the San Joaquin Valley and the San Bernadino
Mountains, it bends in a more westerly direction. This
is the so-called 'Big Bend'. Where this occurs, the
plate motions become rather complex, they no longer

The San Andreas Fault as it passes through the Carrizo Plain, California.

Collapse of a freeway viaduct as a result of the 1994 Northridge earthquake. The officer on the motorcycle sailed off the severed freeway, falling nearly 8 metres to the road below

run passively beside one another, rather they push into each other, compressing the crust into the mountains of southern California, producing faults and earthquakes. At least half of the major earthquakes occur on one or more of the 200 other faults which traverse southern California.

Earthquakes occur when the rocks can no longer stand the pressures that have built up in the crust. Generally speaking, the longer there hasn't been an earthquake, the more likely one is to occur soon. The recurrence time between major quakes in California is between 100 and 150 years, the last great earthquake occurring in 1857 followed by lesser ones in 1872 and 1906. Since 1978 California has experienced a larger number of moderate-strength quakes than in the preceeding quarter of a century, which has led some geologists to predict that a massive quake is due soon. Because of this likelihood – and it has been observed that the crust is opening up in places along the fault line – a major earthquake prediction and mitigation programme has been initiated by the US government.

The history of Californian earthquakes started in 1769, when an expedition led by Gaspar de Portola was violently shaken by a major earthquake while they were camped along the Santa Ana river near the modern town of Olive. California in fact is sliced through by a plethora of faults which break the countryside up into blocks. The San Andreas Fault is the dominant part of this great fault system. It extends for some 900 km through southern California and along the coast ranges of central California. Because of its size and importance to California residents, it has been closely studied.

Its fame spread quickly around the world after the disastrous 1906 earthquake, when 700 people died. We now know that most quakes have their foci about 15 km down, although the fractures themselves almost certainly go deeper, in fact down to the East Pacific Rise below. Within the state the fault zone var-

ies in width from less than 30 m to upwards of a kilometre. Movements along it can be vertical, horizontal or a combination of the two. The total displacement may not be more than 5 or 6 m at a time, but in the past it is clear that crust has experienced periods of considerable vertical displacement that have totalled upwards of 3500 m (witness the scarps of the San Jacinto and San Bernadino Mountains). The 1906 quake was the result of a simple 6.5-m horizontal motion.

It is possible to work out how much the crust has moved by studying offsets of features which have been broken apart. For instance, a stream-bed that crosses the San Andreas Fault near Los Angeles is now offset 90 m from its original course. The sediments in the abandoned stream bed are about 2500 years old. If we assume that movement along the fault cut off the stream bed within this period, then the average rate of slip is 35 mm per year. If we therefore make the assumption that all earthquakes have 4.5 m slip, then quakes should occur on average every 130 years (4500 divided by 35). The reality is that, with this average figure, earthquakes have occurred between 45 and 300 years apart.

The 1906 San Francisco earthquake measured 8.3 on the Richter scale and there was severe damage to homes in the area. After this there was considerable research into building quality and newer buildings were constructed according to quite strict controls in order to minimize the risk of structural failure. In 1971 there was another major quake, this time in the vicinity of San Fernando; this measured 6.6 on the Richter scale and there was still damage even to newer housing. The 1983 Coalinga quake was of the same magnitude, but only affected older buildings. One year later a 6.2 strength earthquake centred on Morgan Hill, again affecting older housing. The 1989 quake was more major, measuring 7.1 on the Richter scale and it was this one which caused the San Francisco freeway viaduct to collapse, despite precautionary building techniques.

The last major quake was the Northridge quake of 1994; this measured 6.7 on the Richter scale. The main shock occurred at a depth of 18 km, involving thrusting along a plane inclined at between 35° and 45° to the south-southwest. Aftershock activity was high, there being 13 aftershocks of magnitude in excess of 4.0 between 18 January and 28 January. Considerable ground collapse occurred, involving the breakage of a gas pipeline and water main. Fires associated with this rupture burned five houses and damaged several more.

From what has been said, it is clear that this is a zone of concentrated natural hazards. A major earthquake is considered well overdue; when it will come, and where, is not entirely certain. However, there is little doubt that, despite improvements in construction and prediction methods, it could be a very frightening experience if it hits any of the cities which lie along the track of the fault zone.

Danger with mud

The small town of Armero is situated in Columbia, South America. On the night of 13 November 1985 most of the 25 000 inhabitants were in bed, but many were afraid of actually sleeping. Why? They had heard ominous rumblings from the ice-capped volcanic mountain of Ruiz, and had seen ash falling lightly over their streets. Then, suddenly and in the dark, the ground shook violently and layer upon layer of mud, rock and boulders, moving with the consistency and ease of wet concrete, slurried over the town, drowning or burying 22 000 men, women and children. Hot mud also washed down the Lagunillas River valley and the air was filled with sulphurous fumes. Only those who had evacuated when the ash had arrived, or those who lived above the river level, survived to tell their tale.

This Andean volcano had a history of previous

eruption, notably in 1595 and in the early part of the nineteenth century. The 1845 eruption had sent mudflows racing down the same Lagunillas River valley, killing 1000 people as it did so.

In a sense, this disaster could have been lessened; officials had been alerted to the fact that the town was built on an old mudflow and that it had been emplaced during the 1845 eruption which killed 1000 of the local inhabitants. They had also been warned that repeated earthquakes and small ash eruptions posed a decided threat to the safety of the villagers. However, no co-ordinated plan of campaign was mounted; the folly of this is now all too plain to see.

Ruiz is a rather broad, flat-topped edifice, capped with an extensive snowfield. Rising 5389 m, it is one of the loftiest volcanoes in this part of the Andean chain. For over one year before the tragic outbreak, earthquake activity had been increasing, probably marking the rise of a new batch of magma into the conduit. What was thrown out of the vents during September 1985 was old rock – pieces of the existing vent sides; there was no fresh magma at this stage. However, volcanic gases were being emitted, pointing to the presence of magma further down. Immediately prior to the cataclysm, vents had been seen spewing out steam and sulphurous gases from high on the mountain. Then, at 3 am on 13 November, small explosions from the summit crater began, and ash fell on the town below. Six hours later hot pumice began jetting upwards, signalling the arrival of new magma. The final eruption was not particularly big, but because of the snowfield surrounding the summit, the hot pumice and ash melted large quanitites of ice and snow, releasing water that mixed with the pyroclastic material to form rapidly-moving mudflows. It was these which swept down on Almero, killing most of its inhabitants.

Mudflows – or *lahars*, as they are often called (an Indonesian word meaning 'flow') – can be extremely dangerous, particularly when they are hot. At Nevado

The havoc created by an early morning mudslide in La Paz, Bolivia.

del Ruiz they moved downslope at velocities upwards of 35 km per hour, stripping off trees and soil and giving rise to 30-m-high waves of cold and hot mud in the river below. Two or three waves came down the mountain in the space of half an hour; the maximum flow rate being estimated at 47000 m^3 per second. This tide of mud and mud-laden water raced downslope into the Lagunillas River.

Most of the original snowfield still remains around the summit of Nevado del Ruiz. Earthquakes and minor explosions are still occurring, yet still no real plans have been made to prevent another similar disaster happening. The inhabitants of this small community wait anxiously to see what the mountain next has in store.

El Chichon

Because it is very remote, the 1982 eruption of the Mexican volcano, El Chichon, never made the kind of press headlines afforded Mount St Helens or Etna. This was in a way surprising, since its effects were far more devastating. An unimpressive mountain, its

cone rose to a height of only 1260 metres.

Before the eruption it had been in repose for about 1000 years; then, at 11.32 am on 28 March 1982, a mild earthquake shook the surrounding area. This was followed immediately by a violent explosion that ejected a column of rock and gas 17 km into the atmosphere. A few deaths resulted. Some locals left their homes; other remained, not wishing to leave their villages.

The following Saturday, 3 April, seismographs recorded over 500 tremors which gradually built up in intensity until a powerful blast shook the region at 7.32 in the evening. Then, at 5.20 the following morning, the major blast occurred. This blew out the core of the volcano, rather as the cork is blown from a gassy champagne bottle, and sent 500 million metric tons of ash hurtling into the upper atmosphere. The blast and the mantle of hot rocks and ash that spread over the surrounding countryside caused not only deaths to man and beast, but widespread devastation.

Atmospheric scientists tell us that El Chichon's volcanic cloud is the largest observed in the northern hemisphere for 70 years; we have to go back to the

(a)

(b)

The eruption cloud of El Chichon, March 1982. (a) Extent of the cloud of fine ash and sulphur dioxide six hours after the eruption. (b) Extent of the same cloud five hours later, as it approached the boundary between the lower atmosphere and the stratosphere. [Photo: NOAA]

1912 eruption of the Alaskan volcano, Katmai, to equal it. Its path around the world was clearly documented by orbiting satellites which photographed its movements. Most of the dust and ash has now settled, but at the time of writing the *Story of the Earth* (1984) there was still enough remaining high in the atmosphere to cut down on the incoming solar radiation. This led to a slight fall in the world's average temperature and may have a longer-term knock-on effect upon the climate.

Geologists do not really know why the eruption occurred at all, for, unlike Mount St Helens, El Chichon does not lie near a plate boundary, nor is it anywhere near other relatively recent volcanoes. However, erupt it did, and although the total volume of debris it ejected was nowhere near as great as its more illustrious American predecessor, the fact that the energy of its blast was directed vertically upward, meant that its cloud could penetrate to great height. This was aided by the large amount of sulphur dioxide the eruption plume contained.

Fortunately for science, the drifting eruption cloud passed over the National Oceanic and Atmospheric Administration's observatory in Hawaii, from which point a laser radar beam recorded it to be more than a hundred times denser than that of Mount St Helens. Further information came from the orbiting satellite, Solar Mesosphere Explorer, which monitored its progress around the world. This revealed that its base lay some 18 km above the Earth's surface, while its top was at about 38 km – far above the world's active weather systems.

The cloud itself contained fine volcanic ash and, surprisingly, crystals of salt that emanated from well beneath the volcano's superstructure. By far the most voluminous constituent, however, was sulphuric acid in droplet form. This forms when sulphur dioxide gas reacts with sunlight and moisture in the upper atmosphere, forming aerosols. These aerosols and the ash particles have the effect of reducing the energy reaching the Earth's surface from the Sun; the global aerosol mass reached a peak near 12 megatons after this eruption. Exactly how this will affect our long-term weather is still uncertain.

Mount Pinatubo and Mount Unzen

Until 1991, Mount Pinatubo – located in the central part of the Phillipine island of Luzon – was considered inactive on account of it having been dormant for 600 years. Then, in June and July of that year, the volcano erupted violently a number of times, spewing millions of tons of ash and pyroclasts to heights of 15 000 m in the air. This was a particularly dangerous situation, firstly since Pinatubo was located only 90 km north of the densely-populated capital, Manila, and secondly was a mere 24 km east of Angeles: extremely close to the strategically important US Clark Air Base.

Locally the ash reached depths of over 3 m, while heavy tropical rainstorms saturated the unstable volcanic ash and triggered fast-moving and extremely dangerous mudslides (lahars). A large proportion of the ejected material entered the Earth's atmosphere and spread around the world in its upper levels.

The initial eruption went on for 20 hours; a series of massive explosions sending a billowing mushroom cloud high into the atmosphere, turning day into night for the Filipino population. Hot ash rained down over a wide area, causing panic and devastation. Plans were made to evacuate the Clark Air Base and there was general concern about people's safety and wellbeing. Gradually over 0.3 m of ash clogged up the airbase's runways.

Concern was not only felt in the Phillipines; at the same time fears were expressed that renewed seismicity and the growth of a huge dome on the flank of Japan's Mount Unzen, heralded a new eruption. Unzen is located on the island of Kyushu, some 850 km southwest of Tokyo. Beginning on 3 June, several

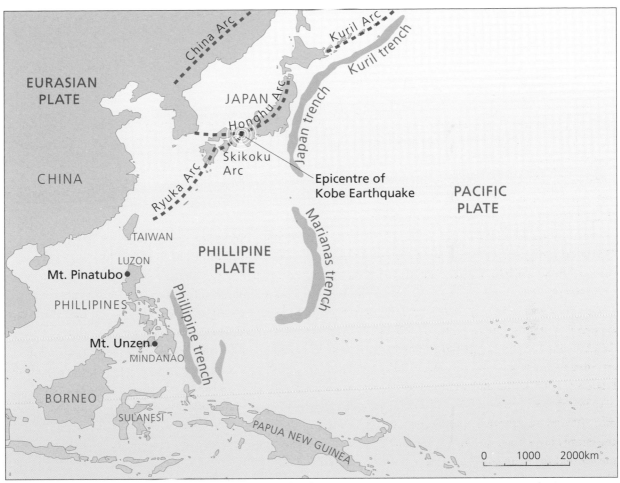

Location map for Mounts Pinatubo and Unzen, and also of the Kobe earthquake.
The plate structure and general geological pattern is also shown.

white-hot avalanches appeared, blanketing Obama, an area 3.5 km southeast of the crater, in ash. Similar avalanches continued during the ensuing week, killing 39 people. Several districts surrounding the volcano began to be evacuated. One housewife was quoted as saying: 'It's the heat – that's what I cannot take.' She was just one of almost 2000 people huddled together in the community halls and schools of the Kyushu seaside town of Shimabara.

Unzen had not been benign in the past; an eruption in 1792 had caused a major landslide and tidal wave that killed 15 000 people. The activity at both volcanoes signalled movements along the plate boundaries of the southwest Pacific and, being typical subduction volcanoes, their style of eruption was guaranteed to be highly explosive, heralding considerable danger to the local populace and wildlife.

The full cycle of activity began in November 1990,

ending a 198-year-period of dormancy. In May 1991 the first mudflows slid down the mountain's flanks and then, on 24 May, the first extremely dangerous pyroclastic flow (glowing avalanche). A series of such flows continued throughout the summer and then, in March 1992, a big mudflow knocked out the Shimabara railway line. Further pyroclastic and mud flows continued sporadically through August 1992 and May, June and July 1993.

The authorities had in some measure prepared for such an event, since a number of erosion-control dams had been constructed. In May 1993 one of the major mud flows was halted by such a barrier, proving its worth. Further mudflows during June 1993 temporarily cut off Shimabara from the outside world. In February 1994 the first pyroclastic flow appeared from the crater's northwest face, spreading downhill in all directions. Following on from this robotic earth-mov-

Mount Pinatubo erupting, 15 June 1991. Heat, smoke and ash caused damage at both the local and global level.

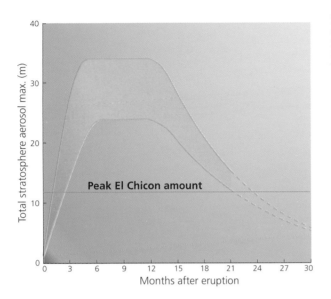

Evolution of the Pinatubo aerosol layer as monitored by the SAGE II instrument aboard the Earth Radiation Budget Satellite. It shows that the level peaked near 30 megatons just after the eruption and started to decline by mid-1992, as significant numbers of particles began to precipitate into the troposphere.

ing equipment was brought in to clear away the volcanic debris.

Eventually, in April 1995, the main Highway 57 reopened to traffic for the first time in two years. Further earth-moving operations were completed. Then, on 12 February 1996, the first pyroclastic flow for over a year descended the mountain. It may be that this will reoccur over a protracted period.

Mount Pinatubo erupted again in August 1992, causing further devastation. Of considerable interest – as it was in the case of El Chichon – was the large volume of micrometre-sized sulphuric acid aerosol particles that were sent up into the Earth's stratosphere. Eventually these were dispersed around the entire globe and had a significant impact on the Earth's radiation budget and the chemistry of the stratosphere. For example, tropical stratospheric temperatures closely following the eruption were incredibly enhanced compared to the 30-year mean, while global temperatures dropped by about 1 °C during the year and a half following. It has been suggested that unusually low ozone levels and high active chlorine levels recorded over the Antarctic during 1991 and 1992, were a spin-off from this eruption.

The evolution of the Pinatubo aerosols has been monitored globally by SAGE II (Stratospheric Aerosol and Gas Experiment II), an instrument mounted on the Earth Radiation Budget Satellite. Global aerosol mass peaked at 30 megatons after the main eruption, and then started to decline until, about one year later, a significant number of particles began to sediment out into the troposphere. By mid-1993, the global aerosol mass had declined to about 12 megatons and by early 1996 had dropped to just over 1 megaton.

The Japan earthquake of 1995

One of the most destructive earthquakes ever to have hit Japan was that which hit Tokyo in 1923. This left 100 000 people dead, 300 000 buildings destroyed and lifted the coast by 2 m. A huge tsunami swept along the coast; the quake measured 8.3 on the Richter scale. Japan experiences something like 5000 tremors every year, Tokyo itself averaging 150. Being a very densely-populated region, such natural phenomena pose a particularly serious problem.

The reason for all this activity is the situation of the islands; they lie just off the mainland of south-

The Kobe earthquake of 1995 caused extensive damage to communications networks. This is the Kobe-Osaka highway, large sections of which were completely collapsed.

east Asia, close to a series of volcanic arcs which have formed along active plate boundaries. Two kinds of quake occur: some take place due to movements associated with the Pacific and Phillipine Plates, where these dip beneath the Eurasian Plate; i.e. they are subduction effects and have deep foci. Others occur at the rigid continental margin of Eurasia, along transform faults. The latter are more frequent and have foci which are relatively close to the Earth's surface.

On 17 January 1995 Japan's nightmare of an urban earthquake hit the town of Kobe, at the southern end of Honshu Island. Kobe has a population of 1.5 million and is the world's sixth largest port. The quake measured 7.2 on the Richter scale and had an epicentre 20 miles south of the city. Its effects shattered the belief that buildings built to new specifications could withstand such a powerful tremor. Motorways were ripped up, buildings collapsed and many fires started. Of the 150 public quays at Kobe Harbour, only 33 survived the earthquake. Businesses and everyday life

were totally disrupted; the situation was so bad that at one point a Japanese gangster syndicate began providing vital relief to earthquake victims, several hundred people having queued up outside the group's headquarters for food, powdered milk, paper nappies and drinking water.

During the event, the ground shifted by around 180 cm in the horizontal direction and 100 cm in the vertical. Seismologists considered it likely that the strength of the earthquake was due to enhancement of the waves at the contact between hard and soft ground, which characterises the location of Kobe city. This is believed to have contributed greatly to the damage experienced by bridges, railway lines and motorway structures. Many of the expressways were completely out of service for months. Over 200 000 people lost their homes, around 250 000 telephone lines were broken and it took about two months to restore gas supplies to the city. It was the worst natural disaster experienced by Japan for many years.

27

What next?

More natural hazards...

As if the Earth had to prove that things are always changing, since writing this book several more major eruptions have occurred. New Zealand's Mount Ruapehu violently erupted during June of 1996, sending ash over a wide area – even curtailing a training session by the Scottish rugby team, located at Rotorua, 120 km away! A year-long ash eruption is still ongoing on the Caribbean island of Montserrat – making this a much less desirable site from which to view the 1998 total eclipse of the Sun than hitherto it had been. Then, during October 1996, a 3-km-high cloud of steam rose above the Vatnajokull glacier in Iceland. This was generated by an unexpected eruption from the sub ice-sheet Grimsvotn crater, situated 120 km east of the capital Reykjavik. The heat sent steam and gas bursting through 600 m of ice, began melting the ice-sheet surrounding the vent and eventually released a huge volume of floodwater, which burst out from beneath the ice, spreading havoc – but fortunately no loss of human life – over a wide area of Iceland's southeast. Thus was repeated an event which took place in 1938.

Less headline-filling was the eruption of Rabaul volcano, on Papua New Guinea. This was active at much the same time as the Iceland vent, sending ash 4000 m into the atmosphere and causing several hundred people to evacuate the town of Rabaul. In September 1994, the New Guinea capital, Port Moresby, was devastated by similar eruptions of the two volcanoes (one of them being Rabaul) which sit on either side of this heavily-populated town. At that time 30 000 people had to be evacuated, five of them dying during the event. Further devastation was inflicted in July 1998 by a tsunami which swept across the Pacific and killed over 3000 coastal-dwelling people.

On 11 November, 1996, the English region adjacent to Land's End was hit by an earthquake measuring 3.8 on the Richter scale. This was the first such event for 15 years and it lasted one minute. The biggest reorded quake occurred in 1757; this measured 4.4 on the Richter scale.

Also during 1996, Australian scientists were busy monitoring a huge iceberg which had broken off the ice shelf of eastern Antarctica and was moving north towards Australia at the rate of 90 km per month. The pieces of this ranged in size from 730 km² to 48 km² in area. It was suggested that this might be a result of global warming.

The Earth's active zones are a constant source of interest, danger and change.

Ozone holes

The Earth's atmospheric envelope is energized by the Sun and allows us to live on the planet. In the distant past the amount of energy generated by the Sun was much greater than now. Before the evolution of plants and the phenomenon of photosynthesis, much more of the harmful ultraviolet radiation penetrated to the planet's surface, and it has been suggested that this may have been the catalyst which enabled plants to evolve from simpler organisms. Today, too much ultraviolet radiation is harmful to the kind of life which now exists.

Earth produces ozone (O_3) which absorbs short wavelength radiation and eradicates roughly 90 per

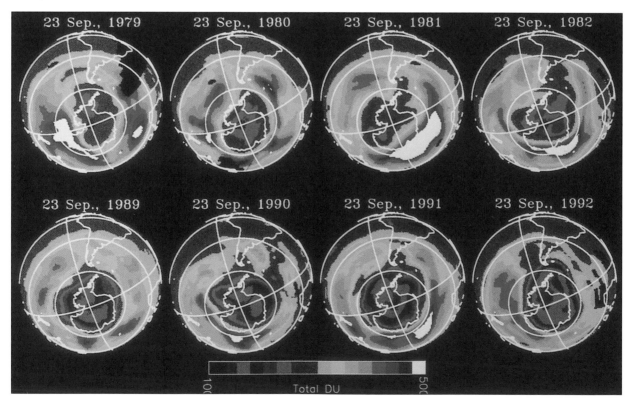

Computer-generated map of the Antarctic ozone hole.

cent of ultraviolet radiation.

During night-time ozone forms by the reaction:

$$O_2 + O + M \rightarrow O_3 + M \text{ (where M = any other molecule)}$$

In other words, the production of ozone requires a three-molecule collision process, which becomes more probable the denser the atmosphere, and requires the presence of both O_2 and O, both produced once photosynthesis began. This suggests that more ozone ought to be produced in low altitudes, however, there is insufficient ultraviolet radiation here to allow photodissociation. The maximum concentration of ozone, therefore, is 30 km above the surface along the Equator, sloping down to 18 km above the poles.

Ozone may be destroyed by photodissociation by ultraviolet radiation during the day, by reaction with nitrous oxide (NO), or by reaction with chlorofluorocarbons (CFCs). The latter two are both generated by Man in this industrial age, and have been blamed for the generation of what are termed 'ozone holes'. The first of these was discovered in 1982 by scientists working in the Antarctic's Halley Bay. They found a reduction in the ozone layer at altitudes of between 12 and 27 km, a finding which was corroborated by later work. Now, since life on Earth is delicately balanced, and any increase in ultraviolet radiation could affect global life cycles and cause skin cancer in humans, this received much alarmist press.

The Antarctic ozone hole turns out to be seasonal, developing only during the southern hemisphere winter and spring (August until early December). It was particularly large during 1987, and spread northwards to south Australia. Since 1989, more than half the ozone over Antarctica has disappeared each spring. In 1996 the hole developed exceptionally rapidly from early September and was of similar size and intensity to that of the previous few years.

Ozone levels are measured in Dobson units. Pre-ozone-hole levels were 300 Dobson units, equivalent to about 100 million tonnes of the precious gas. Satellite data collected by NASA during 1996, revealed a circular area of very low ozone concentration, with an area of 25 million km², covering the Antarctic continent. Readings at the centre of the hole gave 115 Dobson units; this represents a 60 per cent loss on pre-1980 ozone levels. Scientists predicted that in 1997 more than half of the ozone above Antarctica (70 million tonnes) would vanish for the spring season.

Although the most serious deficiency is found over Antarctica, a similar, smaller hole has since been discovered over the Arctic. First spotted in 1988, this appears in the period March–May. Depletion of ozone is not confined to the poles either; US scientists working on Mauna Loa in 1994–5 noted a steep decline in ozone levels in the stratosphere. Although the ozone hole does not directly affect New Zealand, amounts of ozone above this country are about 10 per cent down on 1970s levels. As recently as 1996, meteorologists at Lerwick in the Shetland Isles, reported the lowest levels of ozone recorded in the UK since records began. During 1996, scientists in New Zealand reported an 8.3 million km² ozone hole, equivalent approximately to the land area of the USA. However, despite having high CFC levels, there is more than three times as much ozone over New Zealand than there is over the polar regions. Nevertheless this is about 10 per cent lower than a decade ago.

One of the principal culprits is chlorine; this is used in chlorofluorocarbons which are utilized in industrial processes, such as aerosol propellants and refrigerator coolants. Although these tend to persist mainly in the lower atmosphere, they also slowly migrate into the stratosphere, and it is here that they do the damage, the ultraviolet radiation breaking them down and starting chemical chain reactions which destroy the ozone. Urgent measures needed to be taken, and there have been several international agreements to produce only 'ozone-friendly' aerosols.

Indeed, in 1995, scientists from 21 countries began a series of observations, using helium balloons which can be sent 30 km into the Earth's atmosphere, to monitor ozone levels. The problem is real, and the solution is relatively simple: elimination of the production and usage of chlorofluorocarbons. The strengthened Montreal Protocol of 1992 goes a long way towards attaining this urgent goal.

Pollution

Mankind has contributed to many changes on the Earth. The earliest was the hunting to extinction of certain species – for instance, the woolly mammoth. Subsequently, as the Industrial Age arrived, pollutants have entered the Earth's atmosphere, river sys-tems and oceans. While working conditions in factories may have improved dramatically, the effluents and emissions associated with industry have been allowed to enter the atmospheric and hydrological cycles – with scant regard for the consequences – for many years now.

In northern Europe, for instance, precipitation (rain, sleet and snow) has become between 10 and 80 times more acid during the last 30 years or so. Much of this has emanated from power stations that burn fossil fuels; these produce sulphur and nitrogen compounds which are carried hundred of kilometres up into the atmosphere; later they return to the surface as weak sulphuric and nitric acids.

Europe is not alone in having this problem; the northeastern USA, Russia and Japan are also affected.

Humankind's romance with the internal combustion engine is causing ever-increasing emissions of noxious gases. Modern street scene in northern India.

Disasters and hazards do not always make the headlines. However, city rubbish is a very real health hazard to those living in the poorer parts of cities like New Delhi.

The Scandinavian countries are particularly vulnerable on account of their being underlain by silicic Pre-Cambrian rocks. As a result, acid rainfall draining off the land (much of it brought there by weather systems from other regions of Europe) is not neutralized by carbonate-bearing rocks, and becomes even more acid. Of Sweden's 90 000 medium- to large-sized lakes, 18 000 have become between 10 and 100 times more acid, while 4000 of them can no longer provide suitable habitats for wildlife.

Another pollutant is the offwash from agriculture. Fertilizers, being spread in greater quantities than ever before, and over large and larger areas, also enter the hydrological system. Many of the elements which enter solution, or settle out in rivers and lakes, are harmful to wildlife and may destroy or at very least, weaken, certain elements of the flora. This has a knock-on effect, which also includes the pollution of drinking water supplies.

A great deal is heard these days about traffic emissions: diesel particulates and carbon compounds being two of the worst types. This has become a seri-

ous problem. The amount of carbon dioxide entering the atmosphere has steadily increased over the past 20 years or so, in part coming from traffic and in part from industry. This heavy gas can contribute to a greenhouse effect – something very clearly seen on our neighbour world, Venus, where the greenhouse keeps surface temperatures at roughly the melting point of lead! Carbon dioxide allows solar radiation to enter the Earth's atmosphere, but prevents the reemitted radiation (now at longer wavelengths) from escaping into space. In consequence the heat energy is trapped within the atmosphere, raising overall temperatures. Without suggesting that human activity could produce anything quite as devastating as the Venusian greenhouse effect, it could cause a rise in temperature sufficient to change the climate globally, tipping the balance against at least some species or even genera.

Man's activities are rightly seen as a major contributing force for change in the terrestrial environment. If the destructive trends are to be reversed before wreaking more havoc, then pollution, destruction of

Precipitation map (inches) for the western USA, December 1996.

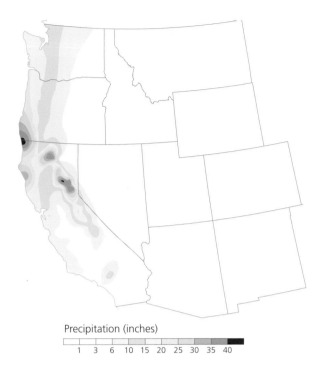

Precipitation (inches)

1   3   6   10 15 20 25 30 35 40

stratospheric ozone and deforestation (which affects the atmospheric balance) are all man-made phenomena which must be tackled internationally, otherwise they will continue to threaten the environment and the organisms which currently thrive in it.

The western United States flooding of winter 1996

The title of this last chapter: 'What Next?' has proved appropriate insofar as, since beginning to write it, details of the extensive flooding that affected much of the western USA have become available. During the month of December 1996, and the first few days of January 1997, tremendous amounts of rain and snow fell over the states of Washington, Oregon, Idaho, California and Nevada. In particular, between Boxing Day 1996 and 3 January 1997, 500 cm of rain fell in some locations, along with unusually high temperatures that generated massive snow melt.

As of 8 January, 1997, in California, 42 of its 58 counties were declared disaster areas. The much-travelled US-50 highway across the Sierra was cut in five places by mudslides and flooding, while State officials estimated at least US$1.8 billion in damage to private and public property. In Nevada, Reno alone estimated US$170 million in damages. The death toll, as of 8 January 1997, stood at 30.

This was the third successive year that the west coast of the USA had been hit by severe floods. The year of 1995 was an El Nino year which experienced warmer than normal sea temperatures in the equatorial region of the eastern Pacific. During 1996, however, there was no evidence to suggest that El Nino conditions existed in the Pacific, indicating that some areas, especially central–northern California, were affected both in El Nino and non-El Nino periods.

Evidently something strange is happening to the Earth's climate. Another snippet of recent information: 1996 was one of the hottest ten years since global records began in 1860, according to the UK meteorological office's records (although, naturally, the UK had some severe cold spells!). This appears connected to a change in wind patterns over the northern hemisphere.

The future

There are, of course, events which are entirely beyond out control. The Earth depends entirely upon the Sun. As we have seen, the causes of ice ages are not known with any real degree of certainty, but the precessional cycle marches inexorably on, and surely will cause climatic change in the future. Currently we appear to be in the midst of an interglacial period; one day another ice age will arrive! Presumably the human species will survive, but there might well have to be a sharp pruning of the population. This, at least, would remove another problem which is facing the Earth.

There are other possible disasters. During reversals of the magnetic field, there may be periods when the field is virtually absent, so that harmful short-wave radiation could penetrate to ground level. There is always the risk of a collision with an Earth-crossing asteroid, which would be unlikely to destroy the planet, but which could alter the climate. Yet all in all, it seems that there is still, barring some hideous man-made disaster, a long period during which the Earth should remain habitable.

Nothing can last for ever. The Sun is using up its hydrogen 'fuel', and eventually the supply of available hydrogen will be exhausted. The Sun will then alter in structure, the core will shrink and heat up, while the surface layers will swell out and cool, so that the Sun will become a red giant star. For a period its energy output will be at least 100 times greater than it now is. Even if the Earth survives, life here cannot. The oceans will boil away; the atmosphere will be stripped off, and the Earth as we know it will have vanished.

Luckily this will not happen for at least 4000 million years yet, and probably considerably longer. Neither is it outrageous to suggest that if humanity has survived until that remote epoch, it may have learned enough to save itself. But speculation here is not only endless, but also pointless. For the moment, we must do everything in our power to ensure that Earth remains suited to our form of life.

Appendices

Man and the Earth in harmony. Ancient Yemeni people built these high-rise homes into the rugged hillsides and they have been upkept ever since. Perched atop ridges of old sea-floor basalts, the local people not only have a spectacular vantage point but also collect precious rainfall in hilltop reservoirs. Manaka, Yemen.

Data for the Earth and planets

	Maximum distance (million km)	Minimum distance (million km)	Orbital period	Rotation period	Inclination of axis	Equatorial diameter (km)	Mean surface temp. (°C)	Mass (Earth = 1)	Volume (Earth = 1)	Density (water = 1)
Mercury	69.7	45.9	87.97d	58.65d	0°	4878	+350	0.055	0.056	5.5
Venus	109	107.4	224.7d	243.16d	178°	12102	+480	0.815	0.86	5.3
Earth	152	147	365.26d	23h56m04s	23.5°	12750	+22	1	1	5.5
Mars	249	207	686.9d	24h37m27s	23.9°	6787	−23	0.107	0.15	3.9
Jupiter	816	741	11.86y	9h50m30s	3.2°	143884	−150	318	1319	1.3
Saturn	1507	1347	29.46y	10h39m	26.7°	120536	−180	95	744	0.7
Uranus	3004	2735	84.01y	17h14m	98°	51118	−214	15	67	1.3
Neptune	4537	4456	164.8y	16h03m	29°	50538	−220	17	57	1.5
Pluto	7375	4425	247.7y	6d09h	88–112°	2445	−230	0.002	<0.01	2.0

The history of life

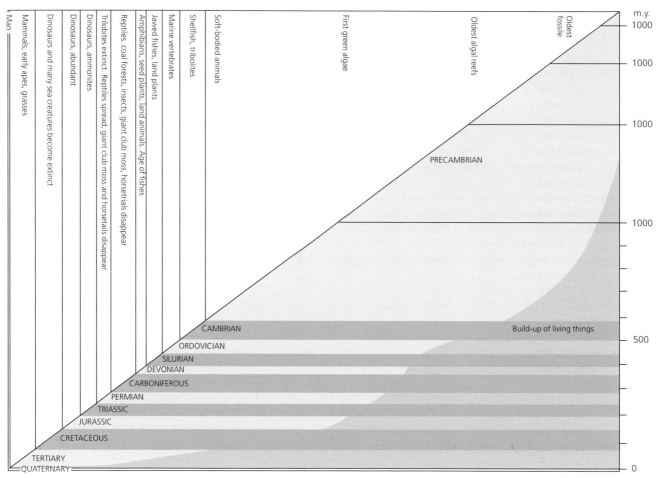

The history of life.

Glossary

Aa: Blocky, fragmented lava.

Anorthosite: A plutonic rock made chiefly from calcium and aluminium silicate. The Moon's highland crust is dominated by this type.

Anticline: A fold which is convex upward. On a map of an eroded anticline, the oldest strata lie at the centre.

Basalt: A fine-grained, dark igneous rock made up principally of plagioclase feldspar and clinopyroxene (augite). It is rich in iron and magnesium but low in silica.

Batholith: A large, irregular mass of igneous plutonic rock, coarse-grained and intruded into the country rocks. Most batholiths are of granitic or dioritic composition.

Breccia: A sedimentary or volcanic rock made up chiefly of large angular fragments.

Butte: A steep-sided, flat-topped remnant of flat-lying strata which have been dissected by erosion.

Caldera: A large volcano–tectonic depression, formed either by explosion, or by collapse of the surface due to magma being withdrawn from below.

Canyon: A deep valley with steep sides, formed by a river.

Conglomerate: A sedimentary rock containing rounded pebbles and boulders.

Continental drift: The movements of continents relative to one another, and driven by motions within the Earth's mantle.

Core: The central part of the Earth below 2900 km. It is made up principally of iron and nickel. The inner part is solid but the outer is liquid.

Craton: A part of a continent which has not been subject to any major deformation for a lengthy period; it may have a thin covering of sedimentary rocks.

Crust: The outermost, brittle, layer of the lithosphere. Continental crust is largely granitic; the oceanic crust is basaltic.

Crystal: Material in which the atoms are regularly arranged to form a rigid network.

Curie Point: The temperature above which a particular magnetic mineral loses its magnetization.

Delta: Fan-shaped accumulation of sediments deposited in a river mouth where it debouches into either the sea or a lake.

Diatom: A single-celled plant with a siliceous framework, found in the surface waters of seas and lakes.

Differentiation: The process of separation of heavier and lighter materials within a planetary body or magma.

Dip: The maximum angle at which the inclination of a stratum differs from the horizontal.

Dolomite: A carbonate mineral made up of calcium, magnesium, carbon and oxygen.

Dune: Mound of wind-blown sand.

Dyke (or dike): A sheet-like body of intrusive igneous rock whose boundaries cut across the bedding planes or other structures in the host rocks.

Eclogite: A metamorphic rock containing pyroxene and garnet, formed only at very high pressure.

Epicentre: The point on the Earth's surface directly above the focus of an earthquake.

Era: A division of geological time. The eras of the Phanerozoic eon are: the Palaeozoic, Mesozoic and Cenozoic.

Erosion: The process by which soil and rock are loosened, broken down and worn away.

Extrusive rock: Volcanic rock, i.e. igneous rock that has emerged at the surface; includes lavas and pyroclastic rocks.

Fault: A fracture in the crust across which there has been relative displacement.

Feldspar: Silicate mineral group whose crystalline structure is a three-dimensional framework of atoms. The most important group of rock-forming silicates.

Fiord (or fjord): Steep-sided coastal inlet formed by glacial overdeepening of a pre-existing valley.

Focus: The point at which an earthquake originates.

Fold: A flexure in rock strata.

Fossil: An impression, cast, mould or track of an animal or plant, left in rocks.

Fumarole: A small volcanic vent sending out gases and hot water, but not lava.

Gabbro: A dark, coarse-grained intrusive igneous rock containing pyroxene and feldspar.

Geomorphology: The study of landforms.

Geosyncline: A major accumulation of sediments with or without volcanic rocks. A largely obsolete term.

Geyser: A hot spring which violently expels hot water and steam. It is normally due to contact between groundwater and magma.

Glacier: A river of ice which flows downhill under its own weight.

Gneiss: A coarse-grained metamorphic rock with a characteristic banded fabric.

Graben: A downthrown block between two normal faults: a rift valley.

Half-life: The time required for exactly one half of a given quantity of a radioactive isotope to decay. It is not affected by external forces or conditions, and forms the basis of radiometric dating techniques.

Horst: An elevated fault block bounded by parallel fractures.

Igneous rock: Rock formed by the solidification of magma.

Index fossil: A fossil of particular value to the stratigrapher since its restricted range allows correlation between strata in different areas.

Intrusion: A body of igneous rock formed from magma which has forced its way towards the surface via fractures.

Isostacy: The equilibrium between crustal blocks of differing thicknesses and the underlying denser mantle layer.

Isotope: One of several forms of one particular element which differ by virtue of having different numbers of neutrons in their nuclei.

Laccolith: A dome-shaped igneous body produced when magma arches up strata.

Lapilli: Fragments of volcanic rock produced when magma is fragmented and thrown into the air by explosive eruption.

Lava: Molten rock (magma) extruded onto the surface, where it may spread out before solidifying as a lava flow.

Limestone: A sedimentary rock made up chiefly of calcium carbonate.

Lithosphere: The outer, rigid, shell of the Earth. It includes the continents and their plates, as well as the oceanic crust. It also includes the uppermost part of the mantle layer – the asthenosphere.

Magma: Molten rock.

Magnetic reversal: A change in the polarity of the Earth's magnetic field.

Mantle: That layer of the Earth between the crust and the central core. It extends from the base of the crust (10–80 km) down to about 2900 km.

Matrix: Fine-grained material in an igneous or sedimentary rock in which are embedded larger crystals or grains.

Mesa: A flat-topped, steep-sided upland block.

Metamorphic rock: Rock formed by alteration of an existing rock by heat or pressure.

Mohorovicic discontinuity (Moho): The seismic boundary between the crust and mantle of the Earth. It lies at between 5 and 45 km below the surface and represents a significant change in seismic velocity.

Moraine: A sedimentary deposit left at the edge or side of a retreating glacier.

Normal fault: A fault in which the fault plane dips towards the downthrown side of the fault.

Ore: A natural deposit from which one or more metals can be economically extracted.

Orogeny: The sum of the tectonic processes in which large belts of the Earth's crust are folded, thrust, metamorphosed and intruded by igneous rocks.

Outcrop: The area over which a rock body reaches the surface.

Pahoehoe: Basaltic lava with a ropy surface.

Palaeogeography: Reconstruction of the Earth's surface in past times.

Palaeomagnetism: The study and interpretation of ancient magnetic fields recorded in rocks, and hence of changes in the planet's magnetic field in past ages.

Palaeontology: The study of organic remains left fossilized in rocks.

Peridotite: A coarse-grained plutonic rock dominated by the minerals olivine and pyroxene. This kind of material comprises the Earth's mantle layer.

Permafrost: Permanently frozen ground or bedrock, of which the surface layer may thaw in summer.

Reverse fault: A fault in which the fault plane dips towards the upthrown side of the fracture.

Sandstone: Sedimentary rock composed of grains 1/16–2 mm in diameter, bound together by a cement. The commonest constituents are quartz, feldspar and small rock fragments.

Schist: Coarse-grained metamorphic rock in which minerals with a platy structure are aligned in parallel planes due to metamorphic recrystallization under stress.

Scoriae: Cinder-like fragments of lava.

Sedimentary rock: Rock formed by the accumulation and cementation of sediment.

Sediments: Unconsolidated particles of minerals and rocks deposited at the Earth's surface by water, wind and ice, or by chemical or biological processes.

Seismology: The study of earthquakes.

Shale: A fine-grained sedimentary rock made up chiefly of clay and silt, and which is laminated.

Shield: A large region of stable, ancient continental rocks.

Stratigraphy: The science of stratified rocks and their interpretation in terms of geological history.

Stratum: A sedimentary bed or layer.

Strike: The angle between true north and the horizontal line contained in any planar feature (such as a stratum).

Strike-slip fault: A fault along which the displacement is predominantly horizontal.

Stromatolite: A fossil form of dome-like structure generally made up chiefly of calcium carbonate, probably precipitated or accumulated by blue-green algae. Stromatolites are particularly important in Pre-Cambrian strata and have a range well back into the Archaean.

Sublimation: The change of state of a substance directly from solid to gas.

Syncline: A flexure in rock in the form of a trough. In an eroded syncline, or on a geological map of one, the younger strata are in the centre.

Tectonics: The study of large-scale structural features of the Earth's crust.

Thrust fault: A reverse fault with a shallow dip plane.

Tillite: A largely unconsolidated sediment produced by glaciation.

Transgression: A rise in sea level relative to the adjacent land, causing formerly exposed areas to be submerged.

Trench: A long, narrow, deep trough in the oceanic crust, characteristically associated with a convergent plate margin.

Tsunami: A 'tidal' wave generated by an earthquake.

Tuff: Fragmental volcanic rock produced by explosive eruption.

Unconformity: A surface which separates an interval of non-deposition or erosion. An angular unconformity indicates a period of folding and erosion between the time of deposition of under- and over-lying strata.

Vein: A deposit of minerals within a rock fracture.

Vesicle: A cavity in an igneous rock formerly occupied by bubbles of gas and formed by decompression of a magma as it rises towards the surface.

Volatiles: Those components of a magma which are readily lost as gases when the pressure on the magma is reduced.

Volcano: Any vent or fissure in the Earth's crust through which magma, gases or pyroclastic material escapes.

Xenolith: A fragment of a pre-existing rock found inside an intrusive body. Such fragments may be incorporated as the melt passes through crustal fractures.

Bibliography

There follows a selection of books into which the reader may care to delve. Some of these may be out of print, but are extremely well written and would be worth obtaining from lending libraries. Many of them are of a general nature and include their own extensive bibliography. I have also included a few articles which may be found useful. Much information can now be clawed from the Internet and with this in mind some World Wide Web addresses have been added at the end of the list. I hope this may prove useful.

Books

Alexander, David. *Natural Disasters*. UCL Press, London, 1993. The classic work on disaster studies, ranging from anthropology through geology to atmospheric science.

Andel, T. van. *Science at Sea*, Freeman, San Francisco, 1981. A non-technical introduction to ocean science, ranging in scope from continental drift to the Law of the Sea.

Andel, T. van. *New Views of an Old Planet*. Cambridge University Press, Cambridge 1994. An excellent overview of recent thinking about the Earth and of global change.

Blong, R. J. *Volcanic Hazards*. Academic Press, New York, 1984. A very fine book dealing with volcanic hazards and including many case studies.

Bolt, T. A. *Inside the Earth*, Freeman, San Francisco, 1982. A well-illustrated text dealing with seismology, earthquakes, geophysics and plate tectonics.

Brown, G. C., Hawkesworth, C. J. and Wilson, R. C. L. *Understanding the Earth*. Cambridge University Press, Cambridge 1992. An Open University-based primer about the Earth and its workings. Excellent book.

Cairns-Smith, A. G. *Seven Clues to the Origin of Life*. Cambridge University Press, Cambridge 1985. A short text dealing with clay minerals and how they may have played a part in the mystery of life's origins.

Cattermole, Peter. *Encyclopaedia of Earth and Other Planets*. Andromenda, Oxford, 1994. A well-illustrated and very readable account of the Earth and how it compares with our planetary neighbours.

Challinor, J. A. *A Dictionary of Geology*, David & Charles, Cardiff and New York, 1978. A much-reprinted volume of great value to geologists of all persuasions.

Decker, R. and Decker B, *Mountains of Fire*. Cambridge University Press, Cambridge 1991. An eminently readable and amply illustrated account of volcanoes.

Dott, R. H. and Batten, R.L. *Evolution of the Earth*. McGraw Hill, New York, 1980. One of their best straightforward texts on historical geology. A lot of detail in some sections. Many maps of continental palaeogeographics.

Emiliani, C. *Planet Earth*. Cambridge University Press, Cambridge 1993. An exhaustive treatment of cosmology, geology and the evolution of life on Earth.

Francis, Peter. *Volcanoes*. Pelican Books, London and New York, 1976. A very readable survey of volcanoes at a very acceptable price.

Harris, S. *Fire Mountains of the West*. Mountain Press, Missoula, Mont., 1987. Popular account of the volcanoes and volcanic areas of the western USA.

Gregory, K.J. (editor) *Earth's Natural Forces*. Andromeda, Oxford, 1990. A superbly-illustrated compendium of facts about a popular topic.

Krafft, M. and Krafft, K. *Volcanoes: Earth's Awakening*. Hammond, Maplewood, NJ, 1980. A superbly illustrated and well-written book by the late French couple who devoted their lives to the first-hand study of volcanoes.

Gross, M. Grant. *Oceanography: A View of the Earth*. Prentice-Hall, Englewood Cliffs, NJ, 1982. A highly successful introductory book that covers all aspects of oceanography, including the interactions between the sea, land, atmosphere and marine life.

Maunder, J. *Dictionary of Global Climate Change*. UCL Press, London, 1994. A comprehensive and authoritive work on this important topic.

Press, F., and Siever, R. *Earth*. Freeman, San Francisco, 1991. An excellent undergraduate text about the Earth. Much used as a course text by universities and colleges.

Stokes, W. Lee *Essentials of Earth History*. Prentice-Hall, Englewood Cliffs, NJ, 1982. A well-illustrated, good, but rather formal treatment about historical geology.

Stommel, H. and Stommel, E. *Volcano Weather*. Newport, R.I., Seven Seas Press, 1983. The story of 1816 – the 'year without a summer'. Invokes the huge eruption of Tambora as a possible cause. Interesting reading.

Wyllie, Peter *The Way the Earth Works*. Wiley, New York, 1976. An easily digestible treatment of plate tectonics and other global processes.

Scientific American books and readers

These reprinted volumes provide excellent surveys of modern research and ideas about all aspects of geology, most of which have been published in *Scientific American* magazine and begin with introductory remarks by a leading scientist in the particular field. They are profusely illustrated and are all published by W. H. Freeman, San Francisco.

Atmospheric Phenomena, 1980.

Continents Adrift and Continents Aground, 1976.

The Dynamic Earth, 1984.
Earthquakes and Volcanoes, 1980.
Extinctions, 1987.
The Fossil Record and Evolution, 1982.
Life in the Sea, 1982.
The Planets, 1983.
The Supercontinent Cycle, 1988.
Volcanoes and the Earth's Interior, 1982.

Web Sites

Alaska Volcano Observatory
    http://www.avo.alaska.edu
American Geophysical Union
    http://earth.agu.org/kosmos/homepage.html
Australian Environmental Resources Information Network
    http://kaos/erin.gov.au/erin.html
Caltech Division of Geological and Planetary Sciences
    http://www.gps.caltech.edu/
Cambridge Earth Sciences
    http://rock.esc.cam.ac.uk/main.html
Earthquakes and Landslides Hazards
    http://gldage.cr.usgs.gov
EROS Data Center
    http://sun1.cr.usgs.gov/eros-home.html
Geological Survey of Canada
    http://agcwww.bio.ns.ca/
Geological Survey of Japan
    http://www.aist.go.jp/GSJ/

Geology – Elysian Fields
    http://www.rcch.com/hotlist/058.htm
GeoWeb
    http://www.pacificnet.net/-gimills/main.
Global Change Research Program
    http://geochange.er.usgs.gov
Global Volcanism and Environmental Systems Simulation
    http://tribeca.ios.com/~dobran
Institute of Geology and Geophysics at Scripps
    http://igpp.ucsd.edu/
NASA EOS IDS Volcanology Team
    http://www.geo.mtu.edu:80/eos/
NASA Solar System Exploration Division
    http://www.hg.nasa.gov/office/solar-system/
National Earthquake Information Center
    http://www.usgs.gov/data/geologic/neic/index.html
Online Resources for Earth Scientists
    http://www.csn.net/-bthoen/ores
United States Geological Survey
    http://www.usgs.gov/
Volcano Images from the Space Shuttle's Radar
    http://southport.jpl.nasa.gov/volcanopic.html
Volcano World News
    http://volcano.und.nodak.edu/vwdocs/vw-news/vw-news.html
World Data System
    http://www.ngdc.noaa.gov/wdcmain.html

Index

Tethys Ocean 121, 123, 163, 176, 177, 179, 184, 191, 195, 200, 202, 206
thecodonts 186
thermosphere 6
Thira 25, 60, 229
thorium 42, 85, 86
Thria (Santorini) 60, 242
Tibetan Plateau 181
tides 8
tillites 133, 140, 141, 162, 165
Torridonian rocks 132
Triassic period 155, 176, 181, 184, 185, 186, 188, 189, 207
trilobites 32, 145, 167
TRM (or thermoremanent magnetization) 67
Troodos Mountains 157
troposphere 5, 78, 259
T-Tauri stage 15, 20
turbidites 125, 131, 135, 162, 207, 216
turbidity current 125, 135, 173, 207

ultravioilet radiation 77, 262
unconformity 90, 91, 93, 110, 134, 270
Unzen, Mount 256, 258, 260
Upper Gondwana Basalts 189
Uralidemobile belt 155
Urals 181
uranium 42, 85, 86, 117, 157
Uranus 3, 12, 23, 267
Usher, James, Archbishop of Armagh 12

Van Allen radiation belts 6, 65

varved deposits 140
Venus 3, 4, 10, 17, 19, 20, 21, 23, 36, 44, 45, 46, 57, 77, 102, 265, 267
vesiculation 46
volatiles 6, 8, 9, 14, 15, 18, 23, 24, 42, 46, 50, 54, 55, 60, 61, 139, 159
volcanism 21, 26, 44, 45, 48, 49, 54, 57, 60, 62, 98, 104, 120, 125, 129, 134, 144, 156, 164, 190, 192, 195, 197, 201, 204, 216, 221, 223, 225, 247
volcanoes 4, 25, 44, 48, 56, 57, 59, 80, 98, 102, 125, 127, 128, 193, 195, 204, 214, 217, 243, 244, 247, 250, 252, 256, 258, 261, 271

Warrumbungles 197
Wasatch Range 223
weather 5, 7, 78, 80, 256, 264
Wegener, Alfred 118, 119, 121
Weizsäcker, Carl von 13
Western Approaches 207
Western USA floods 266
Wilson, J. Tuzo 57, 271
winds 7, 78, 229, 237, 244, 249
Woodward, John 30
Woolfson, Michael 13
Wright, Thomas 12

Yemen 193, 205, 268
Yilgarn Block 116
Yucatan 10, 92

Zagros Mountains 125, 159, 179, 196
Zechstein Sea 174, 176